ROVER
DRIVERS

TECHNICAL TIPS
2nd Edition

ROVER P4
TECHNICAL TIPS
2nd Edition
by
Stan Johnstone

Published by
PB Software
3 Nelson Road
Ashingdon, Rochford
Essex SS4 3EJ

Printed by
Basildon Printing Company Limited
Fleet House, Armstrong Road
South Benfleet, Essex SS7 4FH

A CIP catalogue record of this book
is available from the British Library

ISBN 978-1-909685-05-5

In 2002, when we published the Technical Tips Book, we knew we were on to a winner. However, we did not realise just how successful it would turn out to be. Everyone was talking about it and it became known as Stan's 'Bible' or Stan's little black book, because of its black cover. We had 2,000 produced and when they were all gone, it became very well sought after. Copies even started to appear on E-bay at exorbitant prices!

Due to that success we decided to add on the other technical items that have appeared in Overdrive since, to make it into, hopefully, an even more collectable book.

I trust you enjoy it and it helps to keep your Rover P4 on the road.

Stan Johnstone

Press Release Photo – September 1949

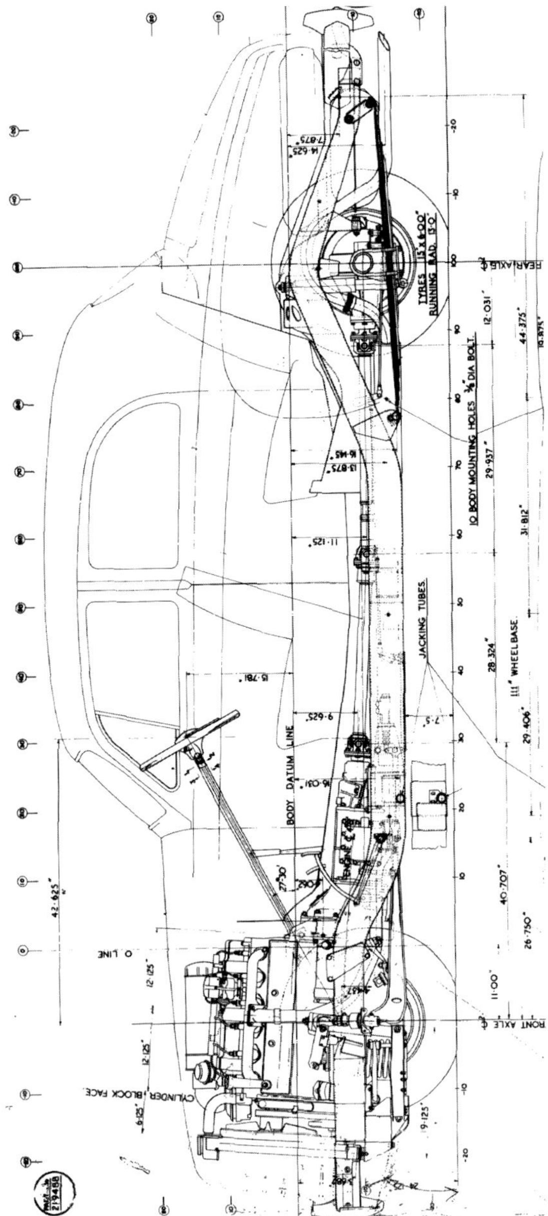
TYRES 15 x 6.00"
RUNNING RAD. 13.0"
10 BODY MOUNTING HOLES 3/8" DIA BOLT.
REAR AXLE ℄
FRONT AXLE ℄
JACKING TUBES
111" WHEELBASE
BODY DATUM LINE
CYLINDER BLOCK FACE
O LINE
ENGINE ℄

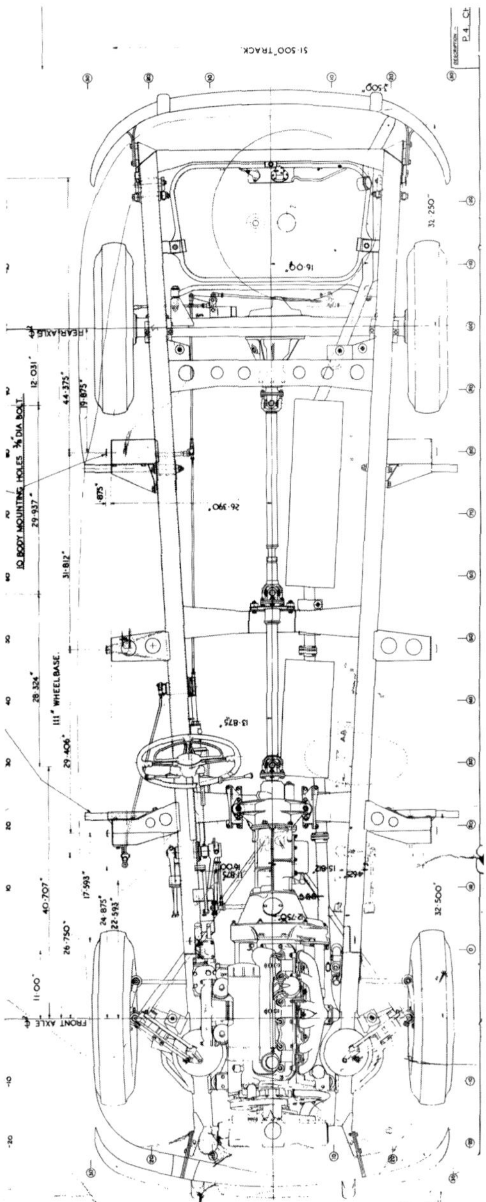
51·500" TRACK.
10 BODY MOUNTING HOLES 3/8" DIA BOLT.
111" WHEELBASE.
REAR AXLE
FRONT AXLE
32·250"
32·500"

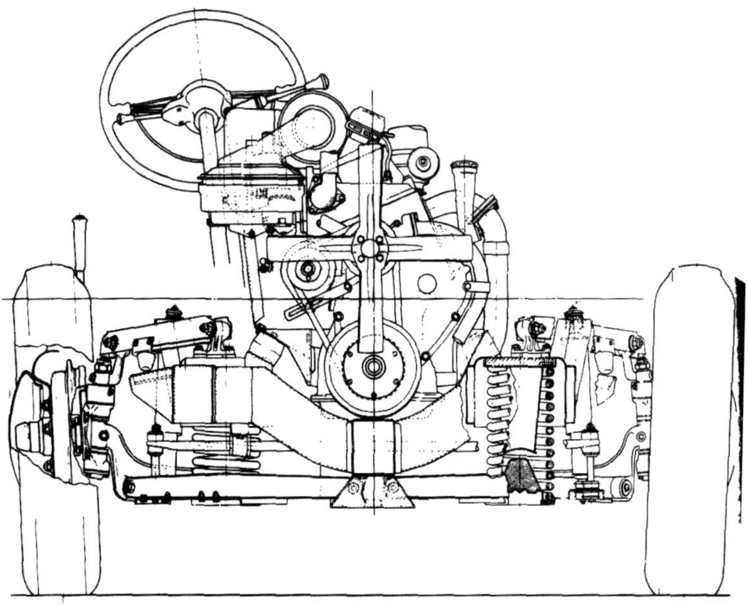

52·00" TRACK.
NORMAL CAMBER 2°

DECARBONISATION AND TOP END OVERHAUL

The following instructions apply to '95', '100' and '110' models in particular, although all six cylinder models are basically similar.

1. Disconnect battery. Set heater controls to "off" position.
2. Remove radiator cap and drain coolant from radiator, and engine block. There is a tap under radiator offside, another one offside of engine block. Use a length of hose pipe on the latter to save soaking the engine compartment.
3. Remove the air silencer/cleaner or oil bath and intake elbow.
4. Remove the flexible air hose to top of carburettor float chamber, the petrol feed pipe and the cold start control cable. Do not lose the small grub screw securing the latter.
5. Disconnect the radiator top hose and the brake air servo pipe at the cylinder head.
6. Disconnect the wiring from the ignition coil, mixture control switch and the low tension lead from the distributor; remove the spark plug leads, distributor cap and ignition leads in one unit. Remove vacuum pipe from distributor, ignition coil and metal cover from above the cooling fan.
7. Take off the alloy heat shield and metal heater pipe from the exhaust manifold. (Note the position of nuts and spacers as this can be forgotten on re-assembly)
8. Remove the top rocker cover and gasket, then the oil feed pipe at rear of cylinder head. Unscrew the thermometer bulb from the cylinder head on '100' models (this may be stubborn, and on no account damage the wire wound tube although this must be gently bent out of the way). On '95' and '100' models pull off the connector.
9. Slacken off the two pinch bolts on the carburettor flexible coupling. Undo the carburettor securing nuts and remove the instrument complete. Do not lose the return spring or choke cable support bracket.
10. Remove the distributor locating bolt and lift out the unit. To ensure correct replacement, scribe a line on the clamp flange and mounting point. Do not slacken the pinch bolt.
11. Undo the three nuts securing the exhaust pipe to the manifold and pull down the pipe. Undo the manifold securing nuts and clamps and remove exhaust manifold.
12. Remove spark plugs.

13. Slacken the inlet tappets right off.
14. Undo the cylinder head securing bolts in the reverse order of assembly, progressively, (see diagram p.3) in order to avoid distortion.
15. Removal of the head is made easier if a wooden handle (old broom handle is suitable) is made to fit over the rocker studs and secured by nuts. The weight is considerable. There will be no difficulty in removing the head but take care to see that the pushrods are left behind in the block or they may fall as the head is lifted.
16. Remove pushrods and retain in their original order (use an inverted cardboard box with six holes).
17. Remove side rocker cover and gasket.
18. On '110' models the rocker gear lifts off as one unit after removal of the securing nuts. On '95' and '100' models, take off the nuts, washers and retaining plates on the thrust blocks, shaft locating screws and end plug. The rocker shafts may be withdrawn using a soft metal drift or hardwood dowel (unless you have extractor Part No. 263749). Note the position of all oil holes for re-assembly, and retain all parts in their original sequences.

Decarbonisation/Valve Removal

19. Remove all carbon from the top of each piston crown using a soft scraper. Work on each piston at top dead centre and wipe off carbon with an oily rag as it is loosened. Stuff some rag into each bore to prevent ingress of dirt.

Inlet Valves

20. Remove valves using a strong spring compressor, do not overtighten. A tap downwards on the valve retaining caps to free the split cones will be necessary. Prise out the split cones and keep in their original pairs. Retain the valves in sequence also. There will be a thick oily deposit on the valve heads and stems; clean this off with a soft scraper. It is essential to remove the old valve guide seals since these harden with age and are responsible for excess oil being burnt. Bc sure to get them out and clean the locating grooves. Grind in the inlet valves using fine grinding paste until a uniform seating is evident on the valve and valve seat, about 1⁄16" - 3⁄32" thick. Excessive grinding should not be necessary. Clean the valves, guides and ports with paraffin. Push new valve seats down the guides, making sure

they seat correctly in their grooves. Insert the clean valves gently (oil the stems first); replace springs, valve caps and split cones using the spring compressor. (Note: on '110' models it is not necessary to remove the inlet manifold for this overhaul.)

Exhaust Valves

21. All the six cylinder engines run "hot" on the side exhaust valves and a good grinding, clean up and even renewal will be required after 50,000 - 70,000 miles, depending on use.
Slacken the tappets right off and use the starting handle to turn the engine over until the compressor can be used on each valve in turn. Use a blob of grease on top of the split cones to keep them from falling down into the sump (if this happens, you are in trouble). Carbon deposits on the valves will be hard and persistent scraping is required. The valve seats are less likely to be in poor condition. Grind in, clean, oil and replace as for the inlet valves. The mating surfaces should be a little wider, however, - about ⅛".

Head Replacement

22. Replace pushrods. Smear a new gasket (marked "This side up") with oil and place on block. Use two metal rods or studs at either end to assist in head location. Make sure the thermostat bypass hose ('95' and '100' models) or rubber sealing ring ('110' models) are correctly positioned. Tighten the cylinder head bolts evenly and in order shown (essential) to the following torques:

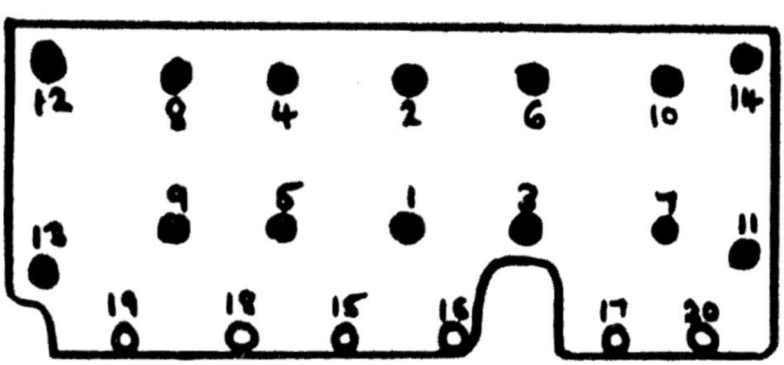

● Bolts 'A' – 50 ft.lbs.
○ Bolts 'B' – 30 ft.lbs.

Set tappets to:	LATE P4		EARLY P4
	inlet	·006"	·008"
	exhaust	·010"	·012"

23. Refit all components in reverse order of removal, but before replacing spark plugs, squirt some clean engine oil into each cylinder and turn over several times on the handle. Make sure the exhaust manifold and faces are clean and use new gaskets, steel face towards manifold. Ignition timing will not need resetting if the distributor is replaced exactly as found (use the scribe marks) since it is impossible to locate this item back to front. Choke operation is dependent on the correct positioning of the grub screw securing point but this too should not have been disturbed.
24. Refill radiator, start engine and check for leaks. Tappets will need resetting when hot. Make sure to re-check head bolts to correct torques after 500 miles.

Richard J. Ginger ***May 1977***

P4 UPHOLSTERY - RESTORATION

If the surface of the leather has been scratched or abrased the affected parts should be treated with coloured lacquer. In some cases, it suffices to "touch-in" the damaged parts with the lacquer, which can be applied by a suitable brush or swab. If it becomes necessary to treat the entire surface of a seat with the coloured lacquer, the directions are as follows:

Having made sure the leather is thoroughly dry (after washing) -

STIR THE LACQUER WELL BEFORE USING.

Pour a small quantity into a shallow container and, dipping a swab of stockinette into the lacquer, apply evenly and sparingly to the leather. If necessary a further coat can be applied after the first has dried.

A second coat can be applied by spray for a more even finish.

Carry out the above renovation in a warm dry shop, or in warm dry weather, this will prevent blushing.

Cleaning Instructions

Waterproof coloured finishes such as "Vaumol" brand – wash the leather thoroughly with warm water and pure non-caustic soap, (Glycerine) or with Connolly's Concentrated Cleaner.

If the Concentrated Cleaner is used, proceed as follows:

Make a solution of one part of cleaner to twelve parts of warm water (the cleaner must not be used in its concentrated form).

Immerse a soft cloth in this solution, wring out the cloth and rub lightly the surface to be cleaned. Avoid flooding and change the surface of the cloth frequently.

In order to remove any residue, repeat the operation with a fresh cloth damped with clean water. Dry well with a soft cloth.

A small brush such as a nail brush (not too hard) can be used to remove dirt which has become ingrained in cushions or the top and corners of the seat.

Connolly Bros ***July 1977***

NOTES ON YOUR P4 CHASSIS

The Rover P4 chassis is not as sound as it may first appear (what blasphemy! - ed.) It is prone to serious rotting in a few places. All these potential trouble spots are at the rear and can be listed as follows:

(1) Under spare wheel carriers, top of chassis.

(2) Inside chassis side wall, both sides, from petrol tank support bar to rear shackle and forward of petrol tank support bar to shock absorber mounting.

(3) Outside chassis side wall from rear jacking post to rear axle bump stop.

(4) Under chassis bump stop mounting - both sides.

(5) End of chassis outer side wall at rear end.

(6) Either side of rear bumper bracket mounting plate on top of chassis - both sides.

Corrosion usually sets in from the outside and if work is undertaken and executed properly, it should be possible to repair the chassis. Although a hole may appear, experience has shown that within ¼" of the edge of the hole on the inside one may still find the original black paint covering completely rust free metal (i.e. inside of box section).

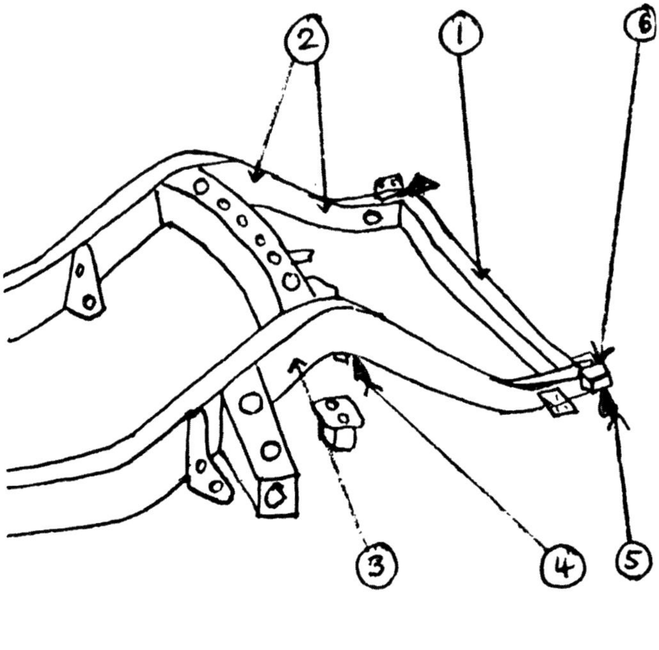

Rust Fighting

Once rust has started to eat away at metal it is almost impossible to stop it unless it is sand blasted. However, the process can be slowed down. Being unable to find an adequately effective neutralizer or rust killer, I prefer to do the following. I prepare thc surface in the normal way and paint a Calcium Plumbate lead based priming paint, which is also a rust inhibitor (BS 3698 type A) called 'Rustodian'. It soaks into the rust infected metal and, after one coat the surface is ready for a top coat. I then apply the old fashioned 2-4 hour lacquer called 'Valspar'. It is now marketed in a new style tin - only one coat is required; it dries within hours leaving a glossy finish second to none.

Replacing Rear Springs

If the original springs are old and rusty (flat) throw them away as they will not contain enough steel to be tampered and re-set. The cost of adding new leaves to give them sufficient strength will be more than the cost of new springs. Difficulty may be encountered when slackening the six shackle self locking nuts, however if you apply the blow-lamp technique you will probably win.

The rear-axle 'U' bolts are difficult to separate from the shock absorber mounting brackets - they cost approx. £2 each. I had to admit defeat and replaced them after removing the old ones with a hacksaw and chisel. Incidentally, they were only removed afterwards by drilling out the mounting brackets. Not mentioned in the manuals, here is a tip worth remembering: The stop brackets bolted onto the chassis can be removed and there is no need to lever the spring into position as the shackles and spring fall into place. The bracket can be fitted with the car standing on two wheels with springs partially compressed by the weight of the car. Another point is that the six bolts - 3 per side - are all different and cannot be interchanged. Everything else should be straightforward.

J H Godfrey ***October 1977***

WINDOW FRAMES

One of the difficulties about writing articles, I have just discovered, is how to start them off. However, I am started now and I hope all the members of Overdrive will forgive any faults they may find, letting me know if they find better ways of doing things, or more particularly, any errors. I am trying to write the article for the complete novice and yet perhaps have a helpful hint even for those who are far more knowledgeable than I. My articles are based on actual work I have recently carried out on my own Rover - a Rover '60' made in 1956 - and will always have a bias towards doing it yourself as cheaply as possible.

One of the problems with the wooden window frames after many years is that the shellac coating put on at the factory becomes chipped and discoloured. A new coat improves the appearance considerably.

Removal of Frame

If you look at Fig. 1 (p.10) you will see that the frame is attached to the metal of the door at six points, (A.B.C.D.E and F). A. and B. are two screws in the front edge of the door when it is open. C.D.E. and F. are half inch self-tapping screws. These are the awkward ones! Firstly, find a screwdriver which fits the slot on the screw head perfectly - these screws are sometimes very tight and it is easy to spoil the head - a long screwdriver, i.e. an electricians' type is helpful because your hand does not cover what you are doing. I had better explain here that finding the screws is a little like looking for a needle in a haystack. Wind down the window, consult Fig. 1 and if you try, say, screw F. (the easiest to find) you will have some idea what you are looking for. The screw is often hidden behind the felt in the bottom or back of the window channel. If it is obviously rusted in to the metal then use one of the usual rust releasers, such as Plus-gas, and leave to soak. Make sure your screwdriver is properly engaged in the screw head, press hard and turn; if very difficult try tightening the screw just a fraction to get the threads free. I was fortunate in that all the screws came out, except three, without too much damage to the slot in the head. On the ones with damage to the slot I was lucky and a larger size of screwdriver blade started them moving and I changed back to the smaller screwdriver to finish them off. If you do damage the slot too badly, it is more than likely that it will be impossible to drill the head off as the screws are very hard. In this case I can only suggest you take out all the other screws, ease the frame away from the door and try to get a small hacksaw blade between the window channel and the door frame to saw through the screw which is shown in Fig. 3 at x. Some damage to the frame and door is bound to result so it is essential you get the screws out the normal way if at all possible. The rear door is just the same as thc front door except that screws A. and B. are in the window channel (G. and H. in Fig. 2). Re-glue any loose joints, if necessary, tightening them with a countersunk screw through the joint but not out the other side. Make sure the frame stays in alignment.

Treating the Frame

The frame is sprayed at the factory with some sort of shellac which gives the wood an even coloured finish, and I have not discovered what product this is. Can anybody help? In view of this I decided to use Sanderson's Translac Clear Polyurethane, obtainable from any D.I.Y. shop. I knew this would bring out the colour and grain of the wood and not produce the same even colouring as original, so any purists will have to forgive me. The finished Translac can, when thoroughly hard, be lightly compounded with cellulose rubbing compound to correct any minor blemishes and improve the polish. Take care when removing the old shellac that you work only in the direction of the grain of the wood and that you do not stress the joints in the frame by leaning or pressing too hard. Work on one section of frame, between joints, at a time, being careful not to overlap the joint as the next section's grain goes in a different way. Scores, scratches and scraping show up like a sore thumb in the finished frame, especially if they go a different way to the grain. I personally, used a coarse sandpaper to score the old finish and then carefully scraped each section with an old but reasonably sharp Stanley blade and found that the old finish came off fairly easily. Any marks or old finish which had been left or unnoticed, show up if you paint the frame with turps. This gives you an idea what colour the finished frame will be. Let the frame dry out and correct any marks, particularly removing old finish as it shows badly under Translac. I had to rub one frame down again! By the way, painting with turps, also removes any dust particles. Give the frame two coats of Translac as per instructions on the tin. You will find the frame has sound-deadening felts stuck on and these can be renewed or replaced if necessary. Treat any rusty brackets with a rust killer after removing them from the frame. Clean with appropriate solvent and then spray the bracket with matt black.

Replacing the Frame

Replacement is common sense; just reverse the removal procedure, but a couple of hints are to offer the frame up to the door and mark the door flange with tape where the brackets slot between the window channel and door. Push frame home carefully to check fit, if joints have needed attention, because the brackets disappear when the frame is pushed home and, at least, you will have some idea whereabouts to look in the window channel for the screw holes. Replace any screws which look even slightly suspect because, if in the future you need to get to your door locks or window winders, the window frames have to be removed before you can remove the upholstery panel on the inside of the door. One more point, remember to lightly smear the screws C. to H. with grease as the post 1959 Rovers had metal doors, not alloy doors like the earlier models.

Do not forget a tin or box for the screws before you start.

FRONT DOOR:
Fig. 1

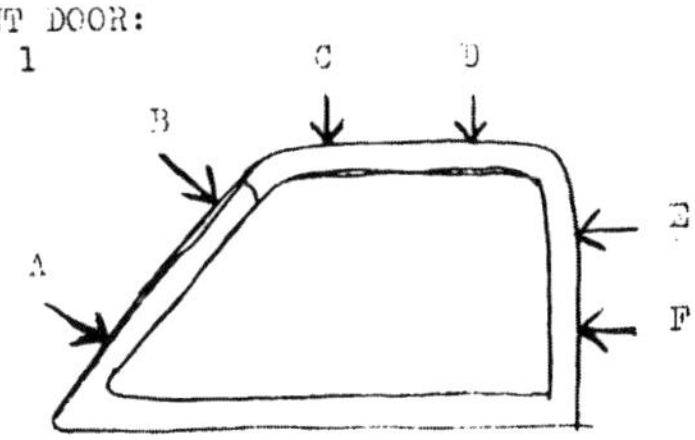

A & B - screws in front of edge of door

C - F - brackets attached to wood frame and slid between metal door frame and window channel

REAR DOOR:
Fig. 2

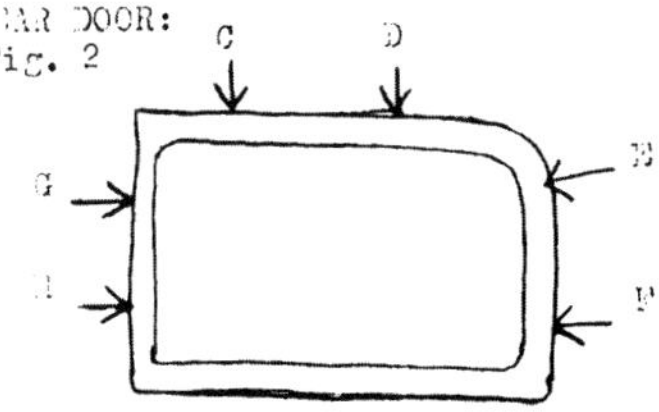

C to H - screws on same place as C to F in fig. 1

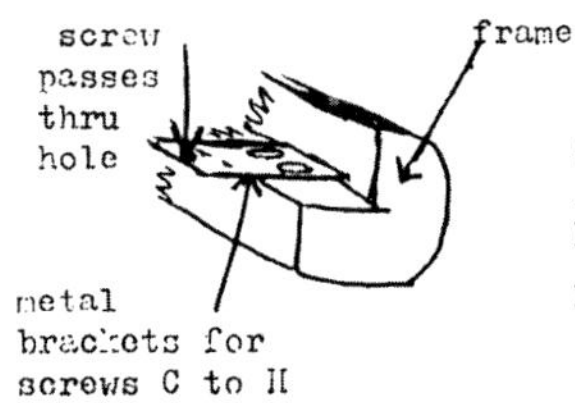

SECTION ACROSS FRAME

Fig. 3

Well, I think I have covered almost everything and, now you have read the articles, if you do decide to do this job, try to incorporate with any repairs to the locks etc. I hope to write an article on the rest of the woodwork, also removal of the door insides soon but, if you are desperate, ring me at Brigg 53351 any evening, Monday to Friday, after about 7p.m. (leave a telephone number if I am out).

Peter M. Wilkinson ***January 1978***

WINDOW FRAMES:

Peter Wilkinson's article in Issue 5 sparked off much written discussion and the reactions of various members follows.

Chris Paling writes: (Chris prefers the orthodox terms and calls them window 'mouldings')

1. I would not advise the use of varnish for when the moulding is heated by the sun and magnified through the glass it will cause the air trapped under the varnish to expand causing blistering or bubbles in the varnish.
2. Rover recommends the following:
   Preparation
   1) Strip the moulding by immersion in cellulose thinners for 5-6 hours or until the original polish is soft.
   2) Scour with wire wool to remove all traces of polish.
   3) Leave to dry for 30 minutes.
   4) Remove surface damage from the moulding with a fine grade sandpaper applied along the grain of the wood.
3. **Staining**
   either
   1) Rub burnt turkey umber mixed with turpentine into the grain of the wood; apply with a rag.
   2) Leave to dry for 30 minutes until the stain powders.
   3) Remove the powder with a clean dry rag.
   or spray with cellulose walnut stain to desired colour and leave to dry for one hour.
4. **Spraying with clear lacquer**
   1) Thin the clear lacquer - not stronger than one part lacquer to two parts of thinners.
   2) Spray with five coats of thinned lacquer allowing 30 minutes drying between coats.
   3) Thin further lacquer to 50% with thinners.
   4) Spray ten to twelve further coats, again allowing 30 minutes drying between coats.
   5) Allow to dry for as long as possible before polishing - at least 48 hours.
5. **Finishing**
   1) Flat to a smooth matt surface using No. 320 wet & dry paper dipped in turps. Care must be taken not to rub through the lacquer on the edges of the moulding.
   2) Polish with a fine rubbing compound using a soft cloth rubbing well in straight lines. Do not use a rotary polishing action.
   3) Finish with a good wax polish applied with a slightly damp cloth.

One of my window frames was apart on the joint and was starting to split along the middle with the grain of the wood. I was advised to put plenty of woodworking adhesive (Evo-stik) on both sides and press together, then sandpaper over the crack when the glue dries. The adhesive becomes clear when hard and the wood-dust settles into the glue. One could not detect the crack afterwards.

I applied clear-gloss varnish mixing in dye (teak) and the result was excellent I thinned the varnish 50/50 with a recommended thinners and applied six coats with a brush rubbing it down with steel wool between applications. The frames look like new now.

Ian Burns writes:

My P4 '100' had been standing for two years with the windows open and therefore the window frames were in a bad way. The front ones had come asunder. After removing the frames, remove the French polish. This can easily be done with a sharp plane blade, using it as a scraper and with the cap iron removed. Also clean up the back of the wood as damp gets in there soaking through the frame causing the finish to be destroyed. Then sand with a fine grade paper. Apply a coat of polyurethane, on the back too, and allow to dry. Sand and apply another two coats. Then finish with wire wool and a good brand of wax polish. This produces an ultra-smooth finish with a semi-gloss shine. The coating of polyurethane on the back wall prevents damp spoiling the finish.

Any pieces that need glueing can be assembled now using "Cascamite" - a waterproof resin glue. If necessary make strips to fit in the slots at the end and when dry excess glue can be chipped off with a sharp chisel. Once replaced in the car the wood can be re-waxed. The best finish is obtained by removing the shine with wire wool and wax, and polishing with a soft rag.

J Butterworth adds that Blackfriar Polyurethane P100 semi-gloss clear varnish gives a preferable subdued finish compared with "Translac". Rubbing down is 95% of the work however, and it is also important to remember to stir the varnish well otherwise it will come out as gloss - the matt particles sit at the bottom:

Another possibly unorthodox approach came through from **Ian Gough**:

While admitting that the right way to set about this job is to remove the frame from the door I managed quite successfully with the frame 'in situ'. The attempts to locate and remove the screws can create more problems and do more damage than the exercise is worth. In addition, if the frame is removed, the joints are often so 'ripe' that it is impossible to prepare the wood for re-varnishing without 'breaking' the frame and then reconstructing the joints.

This is what I do:

I mesh all the adjacent metal and door panel as well as the quarter light. I wind the window right down and then mesh the protruding glass. Trying to remove the varnish coating with sand paper proved to be very expensive both in terms of sand paper and blisters. However, application of sparing amounts of Car Plan paint stripper lifts the top coat very quickly. Provided it is used sparingly this stripper dries as it is scraped off so there is no problem of where and where not to put a sticky, destructive jelly! Instead the dried mixture can be caught in the hand or blown away. Contrary to the instructions on the tin I have found no need to neutralise the surfaces with water before painting.

Once the bulk of the varnish has been removed in this way sanding can remove the rest very quickly. I then coat the wood with 3-4 coats of very thin polyurethane varnish allowing overnight drying between coats followed by two coats of very thin gloss finish polyurethane varnish, meaning 50% thinners and 50% varnish. This brings up the wood grain beautifully and produces a high gloss and durable finish. While allowing the varnish to dry, at least to the 'touch dry' state, do not shut the car door otherwise the furry door piping moults all over your varnish!

Two final notes: when trying to replace a split window frame I was quoted.... wait for it.... £20 by Roverpart of Lewisham and £2 by 'Rover Spares'. This does not appear to be an isolated case....

Returning to the renovation problem - there was an excellent article in the December '77 issue of Thoroughbred and Classic Car.

I.G. ***July 1978***

FRONT SUSPENSION TOP LINK BUSH REPLACEMENT

Jim Hood of Brentwood originally sent in some notes on this and **Peter Rickinson** added his observations too:

Having just replaced very worn suspension bushes on my P4, I thought other members might like to hear how I tackled the job. I might add that I was a little apprehensive of doing this myself, but following warm encouragement from Adrian, the Ilford TC, I had a go.

To ease removal of offending nuts and bolts, these were sprayed generously with releasing oil - WD40 or Plus Gas.

Jack up one side of the car under the front coil spring, preferably with a heavy duty jack, sufficiently to take off all the spring pressure, which then allows full removal of the two plates which make up the top link of the suspension. For safety's sake support the car elsewhere in addition. I used an axle stand under the radius arm.

Having removed the hub cap and wheel, loosen the innermost nuts securing the bolt which passes through the bushes for the mounting of link to chassis. Then do likewise with the nuts for the outer mounting of link to kingpin.

Separate the bushes from the bolts before trying to extract the two bolts. To do so the link must be completely removed, which requires that the two plates making up the link must be separated. This is done by loosening the four nuts that hold on the plate which carries the rubber rebound bump stop. Having so done the two bolts can be extracted and the two spacers removed. The bump stop plate can now be removed and the link plates will be free to be taken off. Now the suspension bushes can be prised from the inner and outer mountings, first one of each pair, then the other carrying the bolt (pin) with it. Clean all parts, including the mountings in which the bushes go. To replace the bushes and re-assemble is simply a reverse of the dismantling technique. Do not use oil or grease anywhere near the bushes themselves as this hastens their breakdown. *(What about rubber grease? – Ed)* Make sure all the nuts and bolts are well tightened and check again that they are tight after a few hours on the road once the new bushes have settled in.

Peter Rickinson writes: From my own observations of doing this job there are some points which I'd like to add:

1. Watch out that you don't stretch the flexible brake hose, which can easily happen when the brake disc and caliper pivot outwards once the top wishbone has been removed.

2. The hole through which the bushes go, on top of the kingpin assembly, can be quite rusty and, rather than forcing the new bushes in, all slavering in grease, it's quite easy to take a round section file and clean the said hole out - then the bushes fit.
3. Waxoyl is an excellent lubricant for these parts.
4. The original bushes are available. I didn't get them or the Metalastik A60 substitute. I used BL Mini rear sub-frame lower trunion bushes. These are the correct diameter but rather shorter than the originals and of softer rubber. I doubt they'll last but they only cost me 11p each...

   Mini engine stabilizer bushes will not fit.

Peter Rickinson ***January 1979***

PETROL PIPES

As petrol pipe corrosion appears to be a common occurrence on P4s, I thought other members might like to hear how I tackled the problem on my 1956 '75'.

On my car the length of pipe over the rear axle was so rusty it just crumbled to pieces. However, the section from the centre cross-member up into the engine compartment was coated with oil and hence well preserved. I decided to re-new from the centre cross-member to the union in the rear wheel-arch.

The materials required are galvanised steel tube (5/16 ins o.d.) and some form of coupler to make the joins. It so happens that 5/16 equals 7.94mm and so 8mm compression-type couplers, as used for micro-bore central heating, are ideal. The pipe I obtained from a motor-factor at £1.48 for 10 feet and the couplers from a builders' merchants at 35p each.

The tube is quite easily bent by hand pressure as the curves involved are not too acute. The most difficult curves to form are those adjacent to the elbow in the rear wheel-arch, as there are three bends close together. Removing the old pipe is no problem. Undo the securing clips where possible. It is not possible to reach the screws holding some of the clips, so these must be prised open to release the pipe. The pipe is then cut with a junior hacksaw and the rear section removed.

Using the original as a pattern, form the three bends at the extreme rear end. Work inwards from the end – it is easier to grip the pipe if you allow a bit extra beyond the first bend which can be cut off later. Form the curve to follow the chassis over the rear axle and then cut off enough pipe to reach about as far as the battery box. Working from the wheel-arch it is possible to thread the pipe between the chassis and the body. It may be necessary to bend the flange on the body outwards slightly to give clearance. When the pipe is in place it can be bent upwards to follow the line of the chassis. This process is actually harder to describe than it is to do. Of course, if you take the body off there would be no problem!

Trim off the extra pipe and connect up to the elbow (preferably with a new olive). Fit the pipe into the clips, positioning it by a bit of judicious bending. Now measure up the almost straight length to join the existing pipe to the new rear section. Join up with the couplers, re-fit the remaining clips and check that the pipe is not strained or rubbing anywhere.

You can now sit back and forget the problem for another 22 years or so – and think of all the petrol you'll be saving.

David Messenger ***January 1979***

FAN MODIFICATION

In recent issues of Overdrive there have been suggestions to remove the fan for quicker warm up and economy. The pros and cons were discussed in brief. The following compromise has proved satisfactory even in heavy traffic.

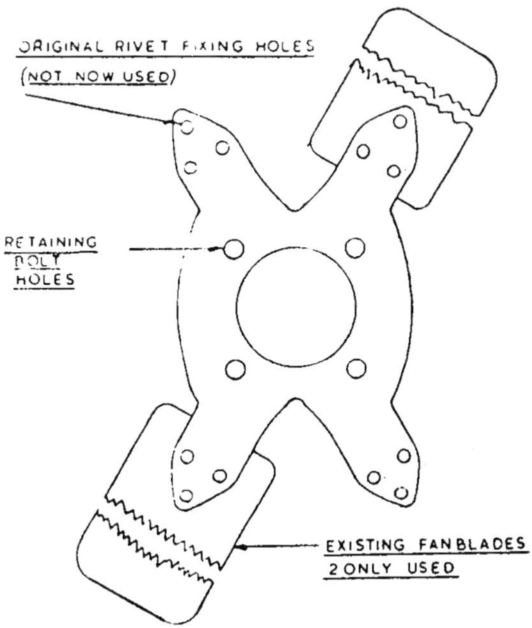

There are four blades on the fan hub, each held in place by three rivets. If one side of the rivets is drilled to the surface of the centre web and then punched through, the blade will come away from the assembly. If this is repeated on the opposite side the result is a two-bladed fan with reasonable balance, capable of keeping the engine cool in heavy traffic.

January 1979

REAR HUB OIL SEALS

It seems to be a widely held view that special equipment is required to remove rear hub oil-seals on P4's.

The Autobook manual states that special tools are needed to service both bearing and oil-seal. This is not the case. The Rover manual is also not clear on this point.

The Rover tool No. 275870 is only required to separate the whole assembly. If everything else is in good order, the seal only may be changed quite simply.

Reference to Fig. E-3 (Rover Manual) shows that the seal is pressed into a sleeve in the housing and contacts the bearing securing sleeve. With the assembly removed from the axle it is simply a matter of pulling out the seal.

"How?" you cry.

The following tip was given to me many years ago by an old Rover storeman.

Drill 2 or 4 equally spaced holes in the outside metal part of the seal. Insert self-tapping screws (NOT countersunk head) into the holes. Take great care to avoid damage. Lever under, or pull on, the screw heads and the seal may be fairly readily extracted.

The new seal can be gently tapped into place with a hammer and drift.

The seal is an ordinary $2\frac{3}{4}$" x $1\frac{7}{8}$" x $\frac{1}{2}$" seal and is easily obtained from a bearing/seal specialist.

I have carried out this job on a number of cars with complete success. Failure of these seals seems to be a common problem, so I don't think that fitting a second-hand unit would be worth all the extra work involved – only to risk another seal failure in the near future.

David Messenger ***April 1979***

EVEN A ROVER SOMETIMES WILL NOT START!

It can happen to the best of us at times, and it happens to some more often than it should!

Firstly, no matter how highly we respect our cars this is no time for emotion; we must apply logic. The stories we have all heard about certain cars being "bad starters" does not apply to the P4. If everything is working as Solihull intended then the car will start on the button.

I will divide this article into two parts: the first looking at the problem of a car that refuses to start after being parked for some time, and second, the roadside breakdown and its primary causes.

If the car refuses to start when it has previously behaved perfectly, it is nobody's fault. On the other hand, if you are aware that the battery won't hold its charge, or the H.T. leads leak and track at the slightest hint of dampness, you cannot really blame anyone but yourself when one cold, wet morning your are late for work and the "longship" refuses to move.
However, Guild members are sympathetic types and even though it may serve you right, let us look at the problem and try to get you on the road again.

When you turn the key or press the starter and the engine doesn't start one of three things must happen: the engine will spin over on the starter but not fire, it won't turn over at all, or it turns over so slowly as to make no difference. If the last is the case, the cause is most likely to be a weak battery and then you will be glad you chose a P4 as the starting handle will probably get you running. Poor connections, either live or earth, may show similar symptoms but they are more likely to produce no reaction from the starter at all, as will a very flat battery. Switch on the headlamps - if they only glow dimly you have a purely electrical fault. If the lights are bright the most likely cause is the starter motor or solenoid.

If the engine spins over but shows no sign of firing (unless you have flooded it, in which case there will be a strong smell of petrol under the bonnet or at the exhaust pipe) your problem is either: no/insufficient sparks or lack of fuel in the cylinders.

The most common cause of breakdown is ignition trouble but with a P4, with the exception of the '80', switching on the ignition and listening for the click of the petrol pump is advised. A sharp tap on the casing will often persuaded a sticking pump to resume service.

Checking for a spark is the next thing to do. Remove a plug lead, expose its metal end and hold it close to the cylinder head whilst pressing the solenoid button, or on later models, getting someone to crank the engine for you (ignition switched on, of course). You should get a sharp, visible and audible spark jumping across the gap. If you do, remove the plugs and check them. If you don't, take your check back a stage. Disconnect the coil H.T. lead from the top of the distributor, hold it close to a good earth, take the distributor cap off and with ignition on, flick the contact breaker points. A very healthy spark (or a belt up your arm) confirms that all is well on the H.T. side upstream of the distributor cap and the low tension side is obviously doing its job well. The problem is now likely to be distributor cap, rotor arm or H.T. leads. It may only be dampness. Remember to check the cap centre contact and look for signs of a hairline crack running from a pickup contact inside the distributor cap. If there is no spark or only a weak one from the coil, then look at the low tension wiring. Start by cleaning the contact breaker points. Ensure everything is dry - a squirt of WD40 or similar. Check the low tension lead from distributor body to coil for breaks and tracking. If all is well in the low tension department you must suspect the coil or ignition switch. Check that the high tension lead is making good contact at the coil. Without any test instruments the only way of telling is substitution of coil or switch with units you know to be good.

Should everything be working correctly on the ignition circuit then lack of fuel to the cylinder must be checked. Check the fuel gauge – don't laugh, most of us do it at least once! Next, disconnect fuel pipe from the carburettor and switch on ignition ('80' owners crank engine). A regular pulse of fuel should come out – please be careful, it is dangerous stuff. Slackening the union and holding a rag round the joint is safer on a hot engine. If the fuel does not appear and there is fuel in the tank, your problem is either faulty pump (replace or repair) or blocked fuel line (clear by blowing or use a foot pump). If fuel is reaching the carburettor, check that the needle isn't stuck by pressing the plunger on S.U. body (sorry '80' owners, you also have a different carburettor) then remove the air cleaner and peer down the venturi to see if fuel is appearing whilst cranking the engine. Make sure that the dashpot lifts and falls freely and that the needle has not dropped – it has been known. '80' owners check for blocked jets.

By now you should be underway again so let us look at the causes of roadside breakdowns. These also invariably take place when it is raining, you are late and miles from the nearest garage or telephone. It would be very handy to be able to fix the trouble on the spot. I am sure we all have ideas about spares to carry and this is obviously a personal matter which each individual resolves in his own way. I once took so many tools and spares, including a battery operated arc-welder, on a camping holiday to Cornwall that the boot would not accommodate the water carriers, sleeping bags and football. The complaints from the children, squashed in the back seat, convinced me that I must compile a list of essentials.

Fan Belt: make sure it is the correct one by fitting the new one and carrying the old one as a spare. I have never used one but I believe that an emergency belt can be a quick, clean answer which does not require tools or the need to get filthy; but fit the proper belt later.

Points: are the most common causes of breakdown. You will also need feeler gauges to set the gap, but a visiting card can be used – and it works.

Spark Plug: not that nasty old one on the garage bench, treat yourself to a new one!

Condenser: perhaps a rare cause of breakdown but cheap and small to carry.

Hoses: even if you call expert help it is doubtful that they will carry P4 hoses. Ideally, replace with new and carry old, sound ones for spares.

Wire: a coil of wire can solve so many problems from electrics to supporting a loose exhaust system.

Tools: I carry too many and will probably continue to do so, but bare essentials are: Phillips and ordinary screwdriver, small electrical screwdriver, pair of pliers and, although I hate them, an adjustable spanner has its uses. If you are fortunate enough to have a complete tool tray under the dash so much the better. Also a can of WD40, or similar, for damp electrics and a roll of PVC electrical tape.

Workshop Manual: essential for data.

Now to the car. You have no doubt sworn and tried vainly to restart it. Switch on the ignition, check petrol gauge and water temperature. It may be the wrong time to mention it, but on later models the dual fuel pump may have stuck through lack of use. Occasional use of reserve – just to keep it free – is a worthwhile dodge, at least you will know for the next time you need reserve. Water troubles are a common cause of breakdown. If the temperature gauge reads high, carefully lift the bonnet and allow the radiator to cool before removing the cap with a rag over the top. I recently saw a girl covered in boiling water and steam on a garage forecourt – she learnt the hard and painful way! Check for split hoses, leaking radiator and loose or broken fan belt. If there is no obvious water loss then suspect the thermostat. Remove it, you will do no harm without it, warming up will simply be much slower.

If the gauges are reading normal then it is worth a glance at the electrical connections – has a battery lead worked loose? Are the ignition connections loose? Especially the Lucar terminals as used on the coil. If you have driven through rain or surface water, are the electrics damp? –WD40 to the rescue!

Spark plugs are unlikely to cause a complete breakdown. One plug may break down – even two – but this would become obvious when driving. My personal choice is N.G.K. plugs. My '110' can eat Champions in 2,000 miles. But I am sure other people would have their own favourite.

If you are still unable to start, then go through the previous steps of checking ignition system and fuel supply, but bear in mind that a faulty starter motor or solenoid will not cause a breakdown. And especially for '80' owners, a blocked jet will rarely cause a total breakdown – usually a large flat spot or loss of power.

By now 90% of you should be under way, the other 10% are looking worried, reaching for the workshop manual or calling the local garage. Your problem is more difficult to diagnose and may need garage equipment to locate the problem. Just one last item I have met twice myself on late model cars. The driving dog on the base of the distributor can work loose and completely disrupt the ignition timing. An infuriating problem as you have plenty of sparks but they are coming at the wrong time, Symptoms usually include back-firing and explosions in the carburettor.

I have compiled a flow chart which should quickly locate the cause of breakdown in most cases, and I suggest keeping a copy of it in the car. Even if you feel competent to deal with a breakdown it is useful to follow a logical step by step guide, if only to reduce the anxiety level. I don't like to admit it but I have stripped down a carburettor only to find I had run out of petrol!! You may find that you have diagnosed the problem but require a spare which is unobtainable at the local garage. This is the time to contact one of the P4 Breakdown Rescue Volunteers who may well have the spare part or be able to locate it quickly.

Best wishes and may you only need to test your breakdown skills on your neighbour's car.

Kevin Bentley ***May 1980***

REVERSING ELECTRICS TO NEGATIVE EARTH

Most owners will not wish to do this, but I play with ham radio and wanted to maintain a morning contact with a station in New Zealand on my way to work. Modern amateur radio gear is quite capable of being used mobile without much worry over the battery coping, but it is all negative earth. The same goes for all cassette players and combined radio-cassettes that are generally available in dual polarity. Old valve radios can be converted from positive to negative earth with a little time and effort.

This is the sequence of operations:

Disconnect the battery and remove it.

Obtain and fit clamp-type battery leads that enable the old leads of restricted length to be extended so that negative can go to positive, and vice versa.

Disconnect the clock and any radio or cassette equipment.

Check the condition of the armoured lead that goes up to the front of the car along the chassis. Repair and insulate, or replace, as necessary.

Bring the battery to the front of the car, and disconnect the dynamo leads. Connect a stout wire between car body (earth) and the negative dynamo connection. Take another stout wire, connect to the positive terminal of the battery, and flash it across the "F" terminal on the dynamo - do not hold the wire on. This procedure serves to repolarise the dynamo coils.

Re-connect the battery with the new reversed leads, and reconnect the dynamo as before.

Start up to check that all works, and that the charging circuit is operational. Of course, the ammeter will show a discharge when charging, and a charge when discharging.

The battery box cover, and a prominent under bonnet position, e.g. the cross member in front of the radiator, should be clearly marked to the effect that "This vehicle is negative earth". One day the services of the AA, RAC or a garage may be required......

In-car entertainments can now be reconnected appropriately, carefully checking the polarity.

The clock poses a problem, as its polarity cannot be changed. The case serves as the earth for the clock lamp, so I decided to dispense with it. The case was then wrapped in several layers of old-fashioned cloth insulating tape, then taking the lead that used to go to the insulated terminal to the earth, and the old earth lead to the insulated terminal. The clock can be refitted and will be securely held by the insulating tape, and is then started in the normal way.

The SW & CB (or + and -) connections of the coil can be reversed, but this may adversely affect the amount of ignition interference heard over the radio. Use whichever suits your circumstances.

Peter Rickinson ***September 1980***

INTERIOR RENOVATION

In previous issues of Overdrive there have been several references to renovation of window frames but little has been said of glove-box lids, corner filler sections and the all wood fascia rails on pre 1959 models.

I have restored completely these parts on my 1954 '60' and the results are excellent.

To start with glove-box lids; I had considered (briefly) doing them in-situ but thought better of it. To remove the lid from hinges it is necessary to unscrew the hinges from the wood of the lid itself. It is impossible to remove the hinges from the lower fascia rail without removing the entire rail. Hinges are concealed under the leathercloth covering to the rear of the lids. The covering is attached at an awkward angle by small tacks which have to be prised out with a sharp screwdriver. Often the heads break off which makes removal of cloth easy but leaves the pins in the wood. Hinges are then unscrewed and it is then easy to extricate the lid from the check strap on the left by sliding the arm to the forward end of the slot where there is an enlarged 'eye' for this purpose.

To remove the corner fillers there is one self tapping screw at the top and a nut behind the stainless steel rail. Take care not to score the veneer on this piece as you slide it down behind the rail.

Working on parts indoors makes the whole thing much easier. Take off all components taking care not to damage any of the wood. The clip for the lid and the runner for the check strap are prone to rust so emery off all traces and coat with clear polyurethane to prevent further corrosion. Take care removing the leather-cloth which is stiffened internally with damp-proof card. I used a smooth knife to gently lever the tacks out of the wood. You may have to replace the thin strips of card that go round the edges as they are usually in pieces when revealed. It's all straight forward if you are careful and use suitable tools for the job.

I used paintstripper to remove varnish from the bare lid. When it's all been removed using a flat bladed knife, wash the surface thoroughly using a sponge. Avoid immersing completely in water in case the wood warps. Allow to dry and then sand to remove all traces of old varnish. Use sandpaper in the direction of grain and finish with the finest paper. Then wipe over with methylated spirits - twice, to make sure it really is clean. Allow to dry, then mix up clear polyurethane 50/50 with white spirit or thinners (N.B. clear polyurethane seems to produce a good colour and brings out the natural grain really well). It is obviously unnecessary to coat the wood under the leather-cloth section so the lid can be rested on this between coats. Allow 6 hours between coats and very lightly sand over the surface with flour paper and then wipe with meths between each coat. Six or seven coats gives a good surface and does not hide the texture and grain. After the final coat, allow to dry under a cover a few inches above the lid so that there is little chance for dust to settle. Leave to dry hard for 24 hours and then for a really good finish wipe gently with T-Cut in the direction of the grain. This will remove any bits on the surface and leave it very smooth. Wipe clean and polish with a wax furniture polish.

Replace all components noting the following:

1. Some leathercloth bits need glueing - use Bostik or UHU and clean the cloth whilst removed with soap and water.
2. To retack the leathercloth (with thin card inserts) use the old tack holes where possible - new tacks where necessary and use a block of wood between hammer and cloth to prevent damage.
3. Don't forget to attach the check strap before the hinges.
4. The leathercloth has to be left unfastened at the back until the hinges are fixed. Then place a screwdriver on each tack and hammer on the screwdriver carefully pulling the lining taut as you do so. It is otherwise impossible to get at the tacks with a hammer.

The 6 coats operation applies to all the wood. I did the fascia rail in-situ and produced good results, but apparently it is easy to remove. And don't forget the tool tray.

Now it is all replaced it looks perfect which is more than I can say about the rest of the interior but - one thing at a time.

Matt White ***May 1981***

GUDGEON PIN REMOVAL - '60', '75', '90' AND '105' MODELS

To remove the piston and con-rod assemblies of all models (barring the '80', '95', '100' and '110' series) It is necessary to extract their gudgeon pins. Typically these are a relatively loose fit and the job gives every appearance of being straightforward. However, a snag arises by virtue of the sloping engine block face greatly restricting your ability to drift out the pins from above via the combustion chambers, the workshop manual specifies the need for a special extracting tool.

An easy DIY method which reduces the risk of damaging the pistons and con-rod bushes is as follows:

1. Unbolt and remove the exhaust manifold. (You have to do this for a proper decoke anyway.)
2. Extract the exhaust valves.
3. Unbolt the big-ends and push the pistons up their bores. (Note that by having earlier removed the exhaust valves you run no risk of inadvertently scoring a piston on a protruding, open exhaust valve.)
4. Remove both circlips from each piston assembly.
5. Cut a length of ½" wooden dowel (about 12" is fine). Pass one end through the exhaust port of the cylinder concerned until it butts against the gudgeon pin. Gently drift out the pin.

Added advantages of this method are that the con-rod remains supported once the pin is out (no nasty clang as it drops onto those beautiful journal surfaces - which you should have bound with masking tape by now anyway!) and a fine degree of control can be exercised.

Keith Coman ***January 1982***

THE P4 BOOT LOCK

The boot handle/lock like the door handles of a P4 is Mazak and prone to severe pitting. If a replacement boot lock is obtained from a breakers yard the first problem is likely to be obtaining the appropriate key. The '95' and '110' models carry the key identification on the serial number plate at the nearside front door, but in any event the lock number is stamped on the steel latch operating bar '8' in the diagram. However, whilst the lock number is shown the series. (FA, FP etc.) is not, but can be determined by trial and error as only a key of the appropriate series will readily enter the lock.

Having obtained a suitable key you may be home and dry. On the other hand you may find that despite liberal doses of penetrating oil it won't budge and dismantling is the only solution.

The same procedure applies whether the lock is seized in the locked or unlocked position. With the assembly the right way up, use a 1/16" drift to drive the pin 1 downwards. The fulcrum tube 2 can then be driven out from right to left. Separate the handle from the mounting bracket 3 to reveal the lock cylinder retaining pin 4 which again can be driven out forcing it from right to left. Do not be tempted to use any force to move the locking bolt 6 at this stage. Inject some penetrating oil into the hole left by the removal of 4, take the handle by one of its ears and swing smartly down onto a soft wooden block so that the point of impact is close to but not on the lock cylinder 5. With luck the lock cylinder will drop out.

If this fails, drive out the dowel 7 and remove the latch operating bar 8. Carefully drill a $\frac{1}{16}$" hole tight against the locking bolt as in fig. 3 and pass a fine drift (or a blunted panel pin) down the hole and using gentle taps force the lock cylinder out. (If you have a $\frac{1}{16}$ long drill you could drill a deeper hole aiming for the shoulder on the lock cylinder.) Having removed the lock cylinder the locking bolt can be safely removed.

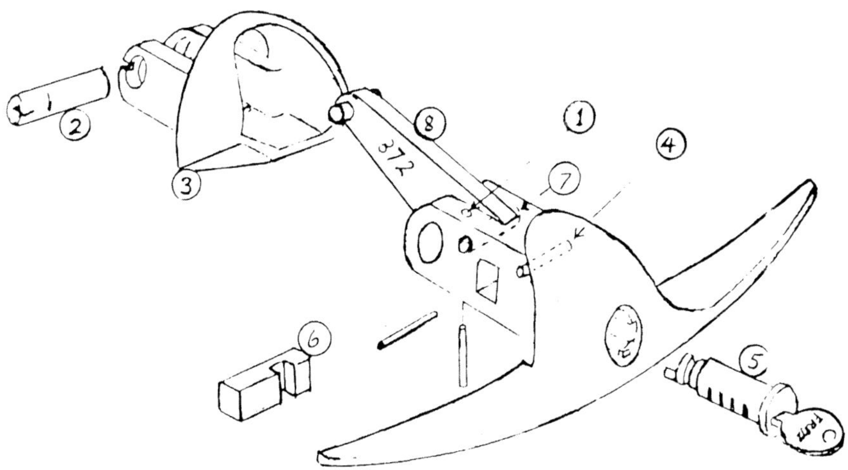

Clean, lubricate and reassemble using either the original locking cylinder or the one from your discarded handle so that the key fits both boot and glove box as it should. Of course an easier way to achieve this might be to remove boot handle and glove box lock from the same vehicle in the breaker's yard.

E.Bromwell ***May 1982***

HOW IS YOUR AUNTIE?

We often find ourselves owning a P4 rather late in the normal car lifespan, and intend to take the car as a regular means of transport rather than to only use the car in fine weather. So we have a late start but doubtless hopes of much pleasurable Rover motoring ahead.

Few P4's, I would think, are bought still with sound coats of paint beneath the car where protection is really needed. On my two P4's, which I take as typical, 100,000 miles plus and 23 and 24 years old, there is quite a considerable amount of original protection now missing. There is surface rust on the underbody and chassis, and from time to time rust has damaged small areas on both body and chassis. Welded new plate repairs have been necessary.

To be realistic we live in a country which is often damp and I doubt that at one time I will have repaired all that needs attention in a given area beneath, have all cleaned thoroughly, degreased then primed and painted with an ideal protective finish. In short, waxoyl or similar, thinned oil, sprayed and brushed is going to be the protection applied and repeated. This is more than either car has had before with previous owners.

Road dirt is scraped and brushed off, paying particular attention to dirt that has built-up in corners, top surfaces and around pipes and fixings (taking care not to cause extra repair work). It is a Forth Bridge type of job if the car is used in all weathers, but done in sections, combined with other jobs it is not too bad. If time is short then an oil spray at the local garage would be better than no action at all being taken.

During a break in the Winter weather or/and at Winter's end, when road salt usage ceases, a garden hose can be used beneath Auntie's wings, paying particular attention to areas around lamps, flange joints and the brake servo reservoir tank. If the car is raised slightly, even one side at a time, the jet can be used to clean quite an amount of the chassis. Whether you wish to flush out the chassis cross-members is up to individual choice; I do, but go for a drive with right and left hand curves afterwards to clear out the water from the drain holes. Two cautions, do not wet too close to engine and take care not to direct jet on front suspension top link housings just cover coil springs - the holes in chassis top will allow suspension bump stop cones beneath to fill. They unfortunately have no drain holes. This is likely the reason for their rusting and eventually collapse.

Existing protective paint and underseal which is removed during cleaning or by the oil/waxoyl used is usually that which is fairly frail.

So far it has been general and there may be members who can speak well of steam cleaning then protection etc.

There are positions on Auntie's chassis where rust can easily get a hold. Mainly the problems stem from a wetted build-up of road dirt. Views extracted from the Rover Parts Catalogue have been used, with certain extras blanked out, to avoid lengthy descriptions of areas I have found in need of attention.

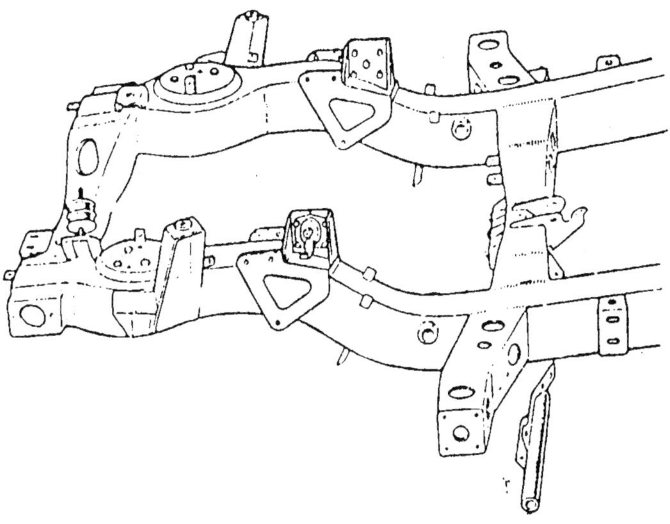

1. **Front Suspension Rebound Stop/Damper Bracket.**
   In this area, hidden to some extent by Inner Wing Panels and Suspension Top Links, road dirt builds on the Rebound Stop and the top surface of the chassis against the Damper Bracket (which doubles as a brake hose mounting). Rust here attacks the chassis and particularly the damper bracket, which on a few cars has been known to partly sever from chassis. This dirt should be cleaned off and waxoyl applied generously.
2. **Steering Idler Mounting (L.H. Side of P4)**
   Steering Box Mounting (R.H. Side of P4)
   The chassis mounting arrangement for above is such that two bolts with distance tubes pass over and close to the chassis. Dirt fills the

space between the bolts, on the chassis top, particularly on the side mounting the idler. Dirt also builds up in the adjacent spaces inside the body mounting brackets and can be seen from beneath the front wings. This dirt if not regularly cleaned out will cause rusting, when wetted, of the chassis. If this cannot be done when dirt is seen, then dribble oil into the dirt, but clean out at a later time.

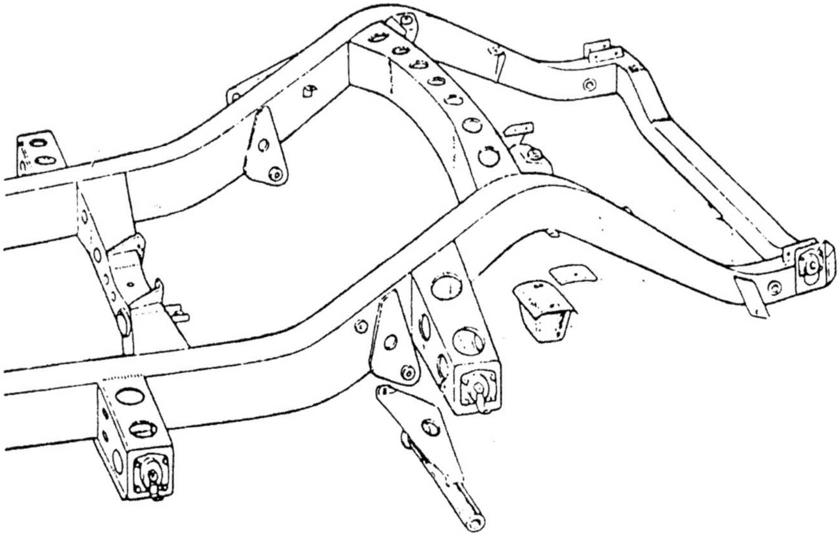

Unfortunately water is to some extent allowed to run from the drainage channels on the top of the front wings diagonally down the bulkhead mountings for inner wings (or wheel arches) and so on to the chassis. Whether this is design or fault on my car I am not sure - any comments?

3. **Attachments to Chassis**

Rear Axle Bump Stops and Packers, Jacking Brackets and items bolted in position and exposed to wetness are disadvantaged unless a sealant is used to prevent ingress of water between the item and chassis (or body). Normally water can remain for long periods to rust the metal; the rust flake expands and thus progressively enlarges the gap, damaging the two metals (distortion can often be seen).

When such items are removed for various reasons, try to ensure that a thick sealant e.g. waterproof sealing compound is used to seal between the surfaces where rust might otherwise attack. Be generous, and after bolting wipe off surplus compound.

Bill Kennedy ***July 1982***

SLUGGISH START ON P4 ROVER CARS

The driving assembly of the starter is held on the shaft by the nut at one end and the sleeve against a collar on the shaft at the other end. The initial end loading is considerable and originally is accurately measured. This is because the heavy spring takes the end loading when the worm is engaging and driving the pinion.

The collar on the shaft is very small and after some use the sleeve rides (or wears) against the collar, until the initial loading of the heavy spring curl the driving end load is transferred to the bearing bush and its housing, so 'pinching' them between the sleeve and the armature. This causes excessive friction. Lubrication is helpful but there is no provision for this.

The cure is:

1) Remove the armature completely.
   (Before doing this it is wise to lift the brushes and springs so locking the brushes clear of the commutator.)
2) Remove the driving assembly complete and separate the end cover from the shaft.
   (The hammer action has probably swelled the bearing bush at one end and there may be a slight burr on the shaft at the collar. You may have to drive the cover from the shaft and correct the bearing bush with a scraper or reamer. But remember that the bearing bush is really worn and loose except where swelled at the end.)
3) Reduce the spigot at 'A' until there is about $^{10}/_{1000}$ inch clearance when the sleeve is hard home with the driving assembly nut tightened.
4) Assemble when the clearance is correct using the distance pieces and thin washers (if any) as before to give, near enough, the original end loading.

Lubricate with a little oil for good measure, and the armature should then spin freely.

My Rover '110' had this done at just over 56,000 miles and has been done again at 163,000 miles. In the interval it has operated perfectly.

September 1982

FRONT WINDOW REGULATORS

The window winders and components inside the doors of a P4 seem particularly prone to corrosion and failure, especially on my 1959 '105'. It got so bad on mine that at one stage the glass disappeared down into the door! The cog on the regulator mechanism on the driver's door had broken so even furious winding on the handle only produced loud clicking noises.

Luckily I recently found a scrap yard with a P4 Rover buried under a pile of other cars – a rather dangerous position – with a VW camper on its roof and a Hillman Imp on top of that. I gingerly entered the seatless shell (the rear door was open). There was little left inside, no useable trim seemed to remain – the interior door trims were in a bad state but had handles still attached. The window regulators were still inside though and within twenty minutes I had removed everything useable from inside the doors. WD40 came in handy here but it didn't help the scuffed knuckles!

Returning home with the prizes I cleaned off all the old grease and rust marks – there was surprisingly little rust on them. I forced as much new grease into the mechanism as possible and did the same with all the moving joints. I decided to waxoyl the whole thing including the glass support channels to try and keep rust attack to a minimum. Originally the channels, which are 'U' shaped in section had thick felt on the inside so that the glass and runners moved up and down smoothly. The best thing to do here is to stick new felt inside after treating it with a suitable rust preventative. Perhaps a carpet manufacturer or upholsterer could supply the right thickness of felt?

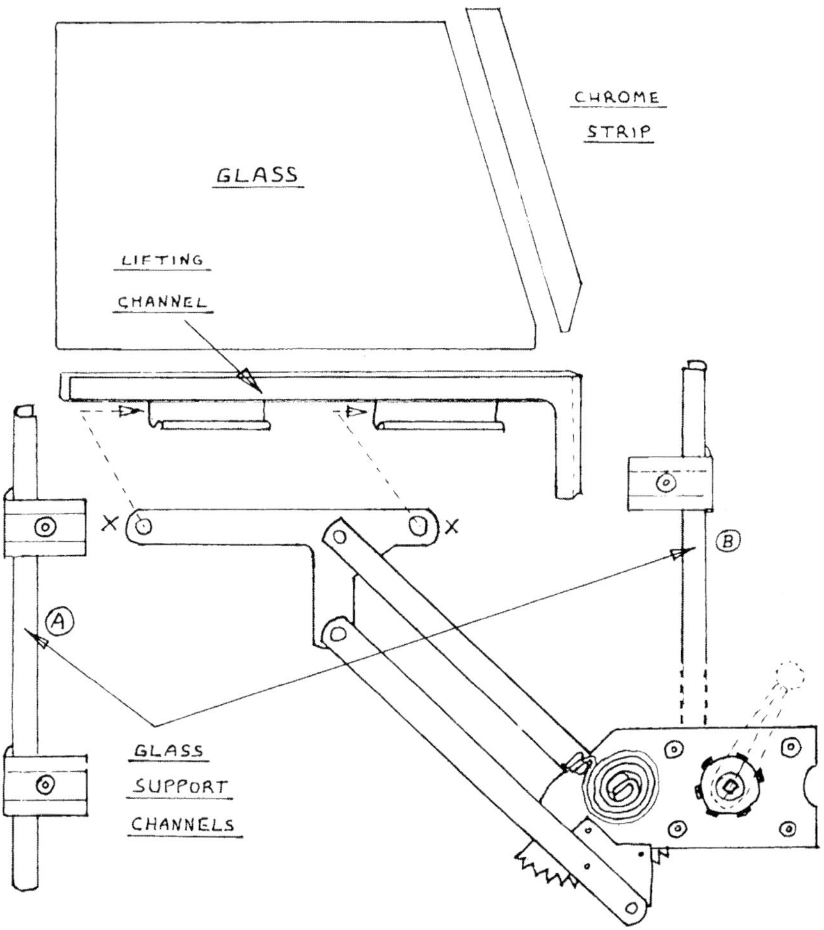

Next comes the job of putting the whole lot back in the door. Fortunately the lifting channels on my car were unrusted and still firmly clamped to the glass. If this is the case you will not have to remove the glass from the door but it must be wedged in the closed position to leave room to work. Of course the draught strips might need replacing, in which case you will have to remove the wooden cappings from the inside of the door and then what's left of the strips before sliding the glass out through the door. The regulator mechanism can be fed through the aperture and attached with the four screws and locking washers. Do not tighten yet. Next, un-wedge the glass and lifting channel and move it forward until the rollers of the regulator arm (marked 'X' in the diagram) can slide onto the brackets as shown. This is particularly fiddly and painful with one arm through the aperture trying to align two rollers on their slides at once! Now the glass is basically attached but can move back and forth unrestricted. The glass support channels control this and only allow vertical movement of the glass. Fit the rear channel first (marked 'A'). The glass itself slides up and down the rear channel but it is the drop arm which slides in the front glass channel (marked 'B'). All the screw holes are elongated to allow adjustment and it is now a matter of trial and error to adjust the positions of the mechanism and support channels to provide the smoothest and most accurate vertical movement. The thing is to get them tight enough not to rattle but easy enough to open and shut. It is important to remember the lock washers on the screws to keep the components from gradually shifting inside the door causing misalignment of the glass.

I can now advertise my car thus: Rover '105' 1959, 192,000 miles, overdrive, black paintwork, grey interior, fully opening front windows.....

Matt White — ***May 1983***

DOOR REJUVENATION (CONTINUED...)

After the hinges and wooden mouldings, what then? P4 doors can intake rainwater at a rate that may make the Water Board envious. The next time you're stuck at the lights in a blustery downpour observe how the water runs down the glass..... and goes inside the door! The water doesn't run off the glass and onto the coachwork.

Firstly it saturates the outer draught strips on either side of the door glass (Part Nos 4 of: 310596 rear and 310597/8 front – two of each). Their capacity to absorb water and thus protect the doors inner mechanisms is limited; and therefore it is probably better not to hose down the windows by way of cleaning or drench the roof with buckets of water. The draught strips dry out readily after a shower – but after any length of time in rain they become saturated and water will proceed further into the door. Thus after 20-30 years they can benefit from renewal since the aluminium part becomes fragile and may crack whilst the felt-fur casing becomes worn and rotten. At worst they may break off and get entangled in the window regulator. Replacement draught strip is available in 2 metre lengths and costs £2.30. Whilst removing the outer length be careful to salvage as many of the securing clips as possible since they are in short supply. The inner strips are riveted to a support and the more substantial rivets on the earlier cars ('55) and more likely to crack upon removal than the more pliable later rivets ('63) which may be used again – although alternatives here could be a little easier. It is possible to replace the draught strips by removing the window mouldings whilst the window glass remains in-situ.

The felt around the flexible glass channelling around the top of the window may also benefit from removal and like the draught strips replacement roll is available.

Meanwhile the window regulators then get a good dousing... this not only controls the rivets making their action difficult but in extreme cases can rot the lifting channel for the glass so the rust accumulated expels the glazing strip and the glass together; this can provide endless hours of fun and entertainment at the wrong time.....

On our wedding day ten years ago the chauffeur desired more fresh air, and got it, as the drivers door glass crashed down into the door. Not wishing to face a draughty 90 mile trip home the groom was seen, later in the day, fishing around discreetly in the door, taking some corrective action. (One day in my life when overalls were not thought necessary – just goes to show...)

All this because of excessive corrosion caused by the ingress of excess water. Later cars, I note, have galvanised parts for the window support channels.

The solution was to fit (after painting the relevant parts with waxoyl) user replacement regulators and a window glass complete with lifting channel from a later car. Certain self-tapping screws can be very difficult to shift – so take care that the screwdriver doesn't slip and damage the softer aluminium panels.

The correct position for the support channels to ensure free window regulation can take hours of fiddling...

After all is reassembled a good spray of waxoyl all around the inside of the door is desirable – this prevents further corrosion (yes even on the aluminium) but has some lubricating properties as well. Where does the water go after that? Ask the owner of any late '95'/'110'! This lack of drainage ensures that the door casings (internal) get damp and may even rot considerably – even alloy doors may corrode just above (and mostly behind) the stainless door mouldings; sufficiently to lift the paint anyway, It is therefore better to keep owners manuals and maps in the glove pocket to prevent the hinges staining the paper, and a screwed up cloth propping the mouth of the map pocket open will speed up the drying out. Should your door casing look 'tired' and 'shabby' yet not too moth-eaten then try giving them a good scrub and wash the carpets thoroughly at the bottom. (Try Fairy Liquid and a soft scrubbing brush). This can have very good results – perhaps raising the state of the door trim – after all, before you became the owner of your P4 how many previous owners would have bothered to give the car a decent spring clean?

This approach also avoids the necessity of buying new carpets, leather upholstery and headlining materials to match newly (unoriginal) retrimmed doors!

Paul Hart ***November 1983***

BONNET RELEASE

With his renewal, Mr Bricket from Fareham, Hants asked about the problem of releasing the bonnet when the inner cable has parted company with the release lever. The workshop manual states:

"Care should be taken that the inner cable is fixed securely to the arm on the bonnet catch plate before shutting the bonnet as it will be found impossible to open the lid other than by operating the cable".

Despite the above warning, Rover designers did not expect the luckless owner to have the ability of Paul Daniels. Instead they thoughtfully provided a small round hole in the grille top support, through which, with the aid of a screwdriver or similar tool the bonnet release lever could be operated.

The hole is to be found immediately to the left of the grille badge, high up behind the grille bars.

Whilst you still have the luxury of a working bonnet release, it will help to raise the bonnet and familiarise yourself with this simple operation.

So who needs Paul Daniels? Well sometimes.

Colin Blowers ***January 1984***

REAR AXLE BREATHER

To prevent pressure building up in the rear axle, a breather valve is fitted to the top of the casing, usually on the offside, although some models may have one valve on each side. If the valve should become blocked or stuck for any reason then pressure that should be allowed to escape into the atmosphere is restricted within the axle and eventually seeks a way out. This condition may cause oil leakage at the seals with possible contamination of brake linings.

Removal and checking of the valve takes only a few minutes with the rear road wheels removed. (Use axle stands for safety)

Firstly clean around the valve housing to prevent any dirt or grit from entering the axle casing and, using a suitable spanner unscrew the valve. Older pattern breathers have a hole in the side of the housing whilst later types are fitted with a shroud.

Clean the outside of the breather and gently shake, the ball inside should be heard to move. Blowing through the bottom screwed opening should be possible with little resistance, if however the ball is stuck on its seating a wash in petrol should free it. This simple task may save your rear brake linings from contamination but will of course not cure leaks from defective oil seals.

Colin Blowers ***March 1984***

FRONT SUSPENSION OVERHAUL

What forced me into such a major undertaking was the discovery, during a routine check, that the R/H road spring had broken what amounted to a turn and a half off the bottom. The broken piece was jammed against the spring support plate and the actual break was quite rusty. So it must have been fractured for some time. I've no idea when it happened as there was no noticeable change in the handling of the car. (The handling was, I must admit, not good!) I had noticed the tyres wearing unevenly but I thought that this was due to worn top link bushes. It probably was but I should think a broken spring didn't help matters. The car did not lean over particularly either.

It took me about fifteen minutes to decide to go the whole hog and dismantle everything except the swivels. (I think that when they need doing I'll let someone else do those!) With the aid of a friend I set about the task. I had soaked all the nuts and bolts the previous day with WD40 and again before I started.

The first job was to jack the car up sufficiently to allow the springs to drop. I placed 12" axle stands under the chassis on both sides where the front outriggers emerge.

The shock absorbers and torsion rod pins must be disconnected from the bottom support plate at this stage. Luckily the shock absorbers simply sheered their bottom bolts very easily. (Note: Add two shock absorbers to the shopping list!) Also remove the central jacking plate.

A trolley jack, by the way, is essential for this job but you don't need a block as described in the manual. I jacked up one side immediately under the bottom spring support plate until the car started to rise as well as the suspension – I knew the spring was then fully compressed. I removed the 'L' section straps on top of the bottom links and slowly undid the nuts securing the bottom plate a little at a time. Eventually a gap appeared between the plate and the bottom links and as I continued the gap became larger but stopped at about ½". Then gingerly I removed all the nuts and bolts leaving two central ones till last. With these out the only thing holding the spring up is the jack and standing well back I used its lowering control from as far away as I could get. Springs are after all a bit lethal if unleashed suddenly! Very slowly the jack lowered and the spring extended and dropped out of the chassis followed by the cone for the bump rubber. I could just remove the jack from under the released spring and manoeuvre the spring and all its assembly out and away from the car. (I think 12" axle stands are the minimum) All this was repeated for the other side and then an inspection of all components. The offending spring was broken in two places. The cones for the bump rubbers were almost rusted through – I could crush the ends with my fingers! The "spring support plate, inner" was non-existent on one side and wafer thin on the other. The holes where the shock absorbers fit were so elongated that they were twice as long as they were wide and the big nut underneath the support assembly was corroded solidly in place. The bottom links came off easily enough and were in good condition, but I could only get one pin out of four free. The top links were O.K. and replacement of the rubber bushes here is quite straightforward with the road spring removed. With it in place it is essential to jack up the wheel till the gap between top links and the bump stop is at its maximum. The radius-arm balls are also easy except that the nuts are difficult to get a spanner to on top of the housing.

Before reassembly of the suspension I had to amass a huge pile of new and second-hand parts. The cones for the bump rubbers had to be partly rebuilt in welded metal (thicker steel than the original was used). This cone is not a precision part and as long as the dimensions are similar to the original there should be no problems. Re-fixing of the rubber to this calls for some invention if new parts are not available. I bonded them in place with epoxy-resin – seems to work as well. In fact the suspension is hardly ever stressed to the point that it will contact its bump stop. I also needed two "spring support plates, inner", two bottom support plates – which incorporate the holes for shock absorber and torsion rod fixings, two rebound rubbers for the top links, two self locking nuts for bottom bump stops, two torsion rod link pins (the replacements were in stainless steel), two shock absorbers and twenty rubber bushes – quite a pile!

Before putting all this lot back every part was cleaned and if necessary painted. Waxoyl is good for every part of the suspension. Rubber bushes and for that matter everything (including the mechanic) was covered before reassembly.

Replacement is straightforward – after replacing the top links the bottom links can go on before the spring is inserted. The cone with bump rubber, the spring and support bush and the bottom support plate have to be manhandled on to the trolley jack and into the spring aperture – accurately so that everything lines up and the ten bolt holes are aligned with each other. You may have to make several attempts at this because as you compress the spring to bring the support plate up to the links it becomes impossible to move the assembly laterally – it sounds complicated to describe but it will be obvious when you do it. Do the nuts up evenly and slowly all round until everything is tight. Make quite sure everything is secure before removing the jack. Fitting of shock absorbers and torsion rod is also easy and straightforward.

The bottom link pin nuts, the nuts securing the radius arm ball housing, the top link bushes and the torsion rod chassis bushes must be finally tightened when the car is in normal riding position. The top link ones are almost impossible to do up with the wheels in place so you can duplicate this position with the wheel off by using a trolley jack under the spring. A few tips – a small blowtorch will aid removal of the bottom link pins if they are difficult to remove – the rubber bushes will soften enough to pull the pins out. A nut splitter is invaluable for removing the large nut under the support bush as it leaves the threads in perfect condition. Necessary tools – a trolley jack, axle stands, ¼ BSF, ⅜ BSF 5/16 and 7/16 spanners and matching sockets, waxoyl.

The special bolts and locknuts in the front of the radius arms may need adjusting so that the wheels do not foul at full steering lock. Wheel alignment will undoubtedly need adjustment as well.

My car was off the road seven days for the job which actually took four days to complete. Total cost of parts including the re-made cones came to £176 which included about £40 on rubber items alone. I must say it was not a difficult job to accomplish but it did take time and patience and I was quite pleased with myself when I drove for the first time – the change in handling and quietness was amazing! After a week or so driving I did go round and check the nuts and bolts for tightness and found that nearly all could take another quarter of a turn. I had to replace the springs as a pair by the way so that the car sat on an even keel. So there it is – don't be too afraid of attempting something that may look complicated; it might be easier than you think.

Matt White — ***May 1984***

OIL FILTER PAPER ELEMENT TYPE

We sell new engines to people who don't change their oil filters – a slogan used in advertisements to make motorists aware of the need to change oil and filters at regular mileages. With little used cars however – and quite a number of P4's fall into this category – the need to change oil and filter becomes necessary at shorter intervals as old oil left in sumps deteriorates and holds harmful acids.

A point worth checking next time you change your oil filter is that the spring and pressure plate are still in-situ at the lower end of the filter bowl. These two items are situated around the centre tube and may sometimes be inadvertently tipped out with the paper filter element. The result is that when the new filter is fitted it simply drops to the bottom of the bowl and performs no useful purpose as oil bypasses the filter to be returned dirty to the sump, rather like having no filter at all. Remember 'We sell new engines…'

Colin Blowers ***November 1984***

LOOK AFTER YOUR 'SWIVELS'!

Regularly check your swivel or king pins, and after a few months you will be able to diagnose their best maintenance routine. For example, if you have renewed or have had the king pins serviced properly, you should only need check the oil reservoir level say, every three months. This is regardless of mileage, as standing still or moving, they can still deteriorate.

If however, you have little or no knowledge of their condition, I suggest you start on the following programme of investigation:

1. Make yourself a king pin dipstick, either using ⅛ inch steel rod, or brazing rod. If these are not easy to obtain 'borrow' (forever) one of your wife's wire coat hangers. You need a length of about 12 inches and using PVC tape, position a mark 6½ inches from one end.
2. Wire-brush the top of pin and air release bolt head on front side of upper swivel housing and remove both bolts.
3. Check level of oil in pins, it should reach the 6½ inch mark. If it is overfilled, it will escape through the air release hole.

NEVER use a grease nipple in the air release hole as this will only damage the seals. If, incidentally, you find you already have leaky seals, you have two options:

(i) Renew seals – which for the work involved really isn't worth doing without replacing pin and bushes as well.

(ii) Alternatively, cut your losses and make a point of replenishing the oil at regular intervals, say every 5-6 weeks. Providing of course, there is not any noticeable play in the bushes.

If on the first check, you find the oil is very murky, or that someone has unknowingly used grease instead of oil, you will have to remove the bottom cover by undoing the tabs and removing the three set bolts. You can then gently flush out the pin with paraffin.

This may not remove all the grease in the bush oilways, but it will certainly get things moving properly again. Replace bottom cover using some form of gasket sealant, then fill with new oil, allowing time for the oil to trickle from air release hold to find own level. An oil can with a flexible spout is the most useful for this job.

Although 140 grade oil is recommended, 20/50 engine oil is better because being thinner, it has more searching properties. Ideally a 50/50 mixture of Molyslip and multigrade engine oil should be used, although of course, Molyslip is a rather expensive commodity.

So generally speaking, a 2-monthly check should keep your swivels in good lubricated order.

Once the level of oil goes below the 6 inch mark, you will then have, what is known in the trade as 'dry balls' – and we all know how dodgy that can be, so be warned!!

Stan Johnstone ***November 1984***

GET THE BEST FROM YOUR HANDBRAKE! (1953-1964)

Depending on the condition of your handbrake system, i.e. the pawl and ratchet at the driver's end through to the slave cylinders at the rear back plate, only you can decide how much work is needed to bring your handbrake to full efficiency. You may only need to do some of the following jobs to get your brakes working correctly.

The pawl is usually the culprit if the handbrake tends to slip. The ratchet, strangely enough, doesn't seem to wear as much as the pawl. A new pawl is reasonably cheap to buy, and quite simple to fit once you have cleaned all the grit and old grease from the mechanism. A pot of paraffin and a ½" brush is ideal for this. Once clean, you will find you can see exactly how much wear has taken place. Re-grease and replace cover when happy that the pawl is working correctly and that there is not too much side play.

Generally speaking, if the linkage is set to the correct angles, an inefficient handbrake is almost certainly caused by seized wheel cylinders (slaves).That is to say, seized on the brake back plate. To remedy this, jack up the car and support on axle stands. Clamp the rear flexible brake pipe and undo the two metal brake pipes, attached to each wheel cylinder. By clamping the brake pipes, bleeding will only be needed to be done to the rear. Release rods from drawlinks. Remove drums and shoes (noting positions of pull off springs). Unscrew the three self locking nuts from the cylinder (note pull off spring positions) and remove the dust cover plate and slave cylinder from the back plate. If the rubber dust covers are dry and both the pistons are moving freely, retain them in the cylinder with a strong elastic band so that the cylinders can be worked on without the pistons falling out.

If the cylinders show any signs of leaks, they will have to be stripped, and if the bores are not corroded, new seals will have to be fitted. Carefully wire brush the drawlink and tappet channel on the cylinder and the face with the studs. Check bleed screw is workable before replacing, they sometimes seize in the cylinder and are much easier to deal with on the bench. Wirebrush the draw links and tappets and clean rollers. Wirebrush inside and out of back plate around the aperture for the wheel cylinder, smear graphite grease or “copperslip” around both sides of the aperture being careful not to use too much. Fit wheel cylinder and dust cover plate to back plate with spring washers and tighten self locking nuts, (preferably new ones) tighten gradually until the wheel cylinder will just slide up and down in its slots using firm hand pressure on the cylinder.

THIS MOVEMENT IS VITAL, IT DOES IN FACT, PERMIT SELF-CENTRING AND THUS A ‘TWO-SHOE’ CONTACT.

Replace the shoes having checked that they have plenty of life left. Screw in bleed screw. Fit drawlinks, tappets and rollers with a sparing amount of graphite grease and secure the cover plate firmly to cylinder. Connect brake rod to drawlink and make sure the balance lever is set at the correct angles. One way of doing this is to copy the angles from the workshop manual on to a piece of cardboard with the axle as the datum line. Whilst both the long and short cross-rods are released from the drawlinks it is worth noting that the relay lever is not seized in its bush and also that there is not too much play in the balance lever bush.

Finally, release clamp from rear brake hose and bleed both cylinders. Adjust shoes until they just start to bind and back off adjuster two clicks.

After all this work take the car down the road, check rear view mirror, heave on the 'Shepherd's crook' and you should hear a loud scream from the back tyres.

Whoever said that P4's have got lousy handbrakes!?????

Stan Johnstone ***January 1985***

HOW TO CLEAN YOUR VOKES AIR SILENCER FILTER ELEMENT

(1954-56; '75' & '90' owners)

1) Follow the procedure in the owners manual for cleaning. (Basically removal of the filter element followed by tapping until the dirt drops out; usually grit and fan-lacerated flies etc.) This should be done every 6,000 miles.
2) Replacements are not available, therefore at 18,000 miles the following may help.
   a) Take a modern washing powder and a tall bucket (over half the height of the felt-gauze element) and make up a suitable washing suds solution in the bucket, full of hot water.
   b) "Stonk" the element vigorously up and down in the bucket, holding each end in turn.
   c) When clean, rinse thoroughly in clean tap water to remove all trace of suds. If you are lucky the filter may just fit the kitchen sink.
   d) Peg out to dry. The felt ends may come adrift slightly but a judicious application of Copydex will correct this and James Taylor will never know. The felt may just clean up to Persil whiteness with care. Washing the felt against the direction of the airflow will also help.
3) Refit when dry, or you will have a problem no damp-start will cure!

Paul Hart ***January 1985***

TIMING TROUBLE

Has anyone an absolutely infallible method of timing an old, partly worn '110' camshaft fitted to a '100' block with a '100' flywheel? It is obvious that with a '100' flywheel and a '110' camshaft the timing marks and the 'time it on the EP mark' syndrome no longer apply due to the '100' and '110' having their timing EP marks in different places. I've played about with a timing disc and pointer and a dial gauge until I've got dials running round inside of my head and figures with them and a '110' which won't do more than 80 flat out.

The cam has still got its performance if I can find out where it is because the original engine used to go really well. I never experienced any trouble in getting it to the ton and beyond before the motor blew up due to a fracas over big end nuts which those of you who also belong to the R.S.R. will doubtless remember. Those of you who have tried to retime an old camshaft will have found that the exhaust peak will have worn sufficiently to prevent this, mine doesn't have a definite peak any more. The top of the cam on No. 1 wanders about over about 5 degrees at peak. I can hear someone saying 'Why didn't he check the timing with a dial gauge, timing disc and pointer before taking the bloody thing to bits'. No go chaps, the motor seized solid when it went round for the last time. I've at present got it timed up at 12 before on the inlet valve but the engine is still as flat as the proverbial pancake. I've also tried 8 before and 10 before without any juicy results, except that it gollops petrol at 8 before and does about 70. What I've not yet tried is taking out all the cam followers and tappets, and checking these for wear although at 55,000 I would have thought this unnecessary. Having done this I could try taking all the opening and closing times of the valves and their present peaks and see what ideas we can come up with between about a thousand of us. What do you think fellas? And no – I am not going to buy a new camshaft just to time the wretched motor.

March 1985 ***Reg Varney***

GIVE THE ENGINE A TREAT

If you are one of those contented people that believe that if their Rover is running well then 'leave it alone', do not read any further! If on the other hand you like to keep a check on your P4, then the following may, I hope, be useful and interesting advice.

Some of the sad cars I have been invited to do work on have broken my heart on lifting their bonnets. Simply because the Rover is such a reliable car, some owners believe they were serviced for life when they left Solihull. I am a firm believer in regular maintenance and think that if you carry out the following jobs and checks, you will be a much happier driver.

Firstly, regardless of your particular car's annual mileage, a change of oil and filter should be carried out every 3,000 miles. The car should be driven for at least fifteen miles before draining the oil into a suitable bowl, remembering that two gallons is a lot of very hot oil. While this is draining, the filter canister can be removed carefully and after removing the old filter and being careful not to lose the spring and washer inside, the filter holder can be washed out with paraffin. The old rubber sealing ring must be removed. This sometimes can be awkward and difficult and if you are not sure you have removed all of the ring, then it is quicker and safer to remove the whole thing from the engine block, and give it a good clean while you are about it. It is not worth leaving the old seal in, as this can invariably lead to leaks, and is not worth the chance. A small electrician's (or come to that a tall electrician's) screwdriver usually gets it moving though.

With all the old oil removed, replace the drain plug. Incidentally, if it is a little worse for wear, Daniel Young can supply new ones for less than a pound. Well worth it, particularly if the wrong spanner or socket has been used and the flats on the head have become rounded off. I had to remove the sump on Gary Edwards' '100' because the previous owners had mutilated the head of the drain plug so much, nothing would get a grip on it. A long job just for an oil change!

With the new filter firmly in place the new oil can be poured into the engine, remembering on starting up that it will take a while for the oil to get round the engine, so keep the revs down until the green light has extinguished. A safer way is to wind the engine with the starting handle a few turns to prime the system.

Whilst on the messy jobs the oil bath air cleaner (except '110' where a new paper element should be fitted) should be removed and after removing the wire gauze, the bowl should be cleaned and refilled with new engine oil, approx. 1 pint, to the the arrow marked on the side. The gauze should be washed in petrol or paraffin and shaken dry and replaced inside the bowl, which should then be screwed back on to its mounting.

Next a new set of points should be fitted to the distributor and on setting the gap (.015") the ignition timing should be checked. I find that a static check is the most satisfactory way and have found it is more reliable than using a Strobe light. I use a 12 volt inspection light with two crocodile clips but even an old spot light will do the job. Having checked the timing setting for your particular model, attach one lead to the low tension connection on the side of the distributor, and the other to a good earth i.e. engine mounting bolt.

Turn the engine with the crank handle (removing the spark plugs first will facilitate this) until the rotor arm is pointing to No. 1 spark plug, i.e. in a straight line with the engine. With the ignition switched on, turn the handle very slowly until the light just illuminates. Having swivelled back the flywheel cover plate (driver's side on all models except '105R' which is on passenger side), check to see where the firing point is in relation to the pointer on the flywheel cover plate. Usually after just replacing the contact breaker points a small adjustment can be made using the vernier nut on the distributor body. If more adjustment is needed the pinch bolt will have to be slackened on the clamping plate at the base of the distributor and the whole thing rotated slightly until the light comes on at the correct setting on the flywheel. If the light illuminates before the correct mark in degrees has reached the pointer, the timing is too far advanced or conversely if the light comes on after the timing mark, the ignition is retarded.

Adjust until happy that the timing is correct, lock up tightly if clamping plate has been slackened and replace timing cover. Remove leads, replace cap and switch off engine.

While the plugs are removed, it is a convenient time to adjust the exhaust tappets if needed. These should be done with the engine stone cold, whilst the inlets must be adjusted with the engine at full running temperature.

The rule of seven should be employed i.e. With No. 6 valve fully open No. 1 valve should be checked and adjusted as necessary. No. 2 valve should be set with No. 5 valve fully open, and so on. Hence the two valves being checked will always add up to seven.

Time should be taken over tappet settings, as they are vital to the well-being of the engine. If anything, it is better to have slightly bigger gaps than being on the tight side, as too small a gap could result in a burnt valve. When happy with the exhaust tappet settings replace the side cover, preferably with a new gasket.

If the spark plugs haven't done more than 12,000 miles they can be cleaned and regapped to .030 thou. and replaced. Otherwise fit new ones but don't forget to gap correctly before fitting.

The car can then be taken for a good run and returned fully warmed to adjust the inlet valves. Again using the rule of seven.

Finally, the two oil breathers, one on the top rocker box and one on the side rocker cover should be removed and washed in paraffin.

This may all seem a bit basic for the average mechanics amongst us, but I really intended this advice for the 'do-it-yourselfers', who, I am sure, are quite capable of maintaining their own P4's. So just to recap:

Oil change and new filter every 3,000 miles. Points renewed generally speaking every 6,000 miles and plugs, oil bath (paper element on air cleaner of '110' changed every 18,000 miles) and breathers changed or cleaned, respectively every 10-12,000 miles.

Stan Johnstone ***March 1985***

LOOKED AT YOUR TOP LINKS LATELY?

One of the most hard-working parts of the P4 front suspension is the top link rubber bushes. This is particularly so if the car is run for many miles with old and inefficient shock absorbers. MOT examiners like to get their long and mean levers under the links and charge you the earth to replace the bushes for they know that, basically, it is a straight-forward job to do. So don't be caught out, check yours and if necessary replace them yourself and you will be surprised how simple it is to do.

Tell-tale signs of wear are a ragged edge round the bushes where they are squeezed between the links and the top anchor. Also with the car jacked up at the centre jacking plate, and the wheels removed, a lever can be used between the links and the chassis to determine any excessive movement. If you do decide to tackle this job, make sure that you can obtain the eight new metalastic bushes (4 each side) before starting the work. Not wishing to provoke any panic buying, but Daniel Young was commenting recently that stocks were getting low and that he and our other friends in the spares outlets were negotiating the possibility of re-production of new bushes. However, enough of my intros, let's get on with the job!

Working on one side at a time, jack up the car safely preferably using a trolley jack. Remove wheel and with a bottle jack under the outer end of the bottom links, raise the suspension until the rebound rubber backing plate is 2½ inches away from the chassis extension. This position is also correct for re-assembly and the top link retaining nuts should be finally tightened in this position.

Undo the two ¼ inch BSF bolts securing the rebound rubber plate. Incidentally, the plate is usually all that is there as the rebound rubber is invariably missing, sometime in the late sixties, I would think. Don't worry too much about the absence of these rebound rubbers, as unless you use the M1 a lot they will rarely be needed.

Remove the bolts, distance tubes and plate and loosen the ⅜ inch BSF self-locking nuts on the link pins. The rearward facing nut at the inner end of the top links can be a bit awkward as there isn't much room for movement. If this nut proves to be too stubborn the other 3 nuts and washers can be removed with the front link and rubber and wriggled off, still attached to the link, it can then be removed with the help of a vice. Another problem that can arise is that the metal sleeve inside the metalastic bush (hence the name) can sometimes become rusted to the pin, but with the aid of the vice again and by protecting the thread with the nut screwed back on, it can usually be persuaded to move with a copper faced mallet. If the pins are badly corroded, they will have to be replaced, either with serviceable secondhand ones or new ones, which are still available through the usual channels. Clean the pins thoroughly with a wire brush and emery cloth and smooth out the surface inside the top anchor housing with a half round file.

This preparation is important as it gives a longer life to the new bushes. Before re-assembly, smear Copper slip on to the pins, and bushes. This will not only aid the assembly but will leave a lasting lubricant on the bushes and again lengthen their life.

Systematically tighten the four ⅜ inch nuts and with the 2½ inch gap as described earlier give them a final tighten. Replace the rebound rubber plate, distance tubes and ¼ inch BSF nuts and bolts. Again it must be stressed that the suspension assembly must be raised to give the 2½ inch gap already mentioned. This will ensure that no extra tension or pre-load is put on to the new bushes.

Well, apart from putting the wheel back on and removing the jacks, that's one side done. Not too bad, was it? Much more rewarding and cheaper than letting that MOT garage do it. Generally speaking they are only used to Escorts, Metros and Yugos anyway!

Stan Johnstone ***May 1985***

COLD START WARNING LIGHT CONTROL

Some P4 owners may have had the same experience as I had when I bought my latest P4. When the cold start knob was pulled out the orange warning light immediately came on, indicating some kind of short in the circuit. Assuming that the fault lay in the cylinder head sender unit, I set about acquiring a replacement from a scrapped P4.

I decided to test the 'new' unit before fitting it and found the same problem existed. I was about to throw it away in frustration when my curiosity got the better of me and I decided to see what the unit contained.

The top plate is made from an aluminium alloy casting and the 'bulb insert' is copper. I separated the two parts with the help of a Stanley knife and revealed a mass of rusted metal within. This was obvious indication of the cause of the problem and also showed that the unit was beyond repair.

I thought that the unit in my own car might be in a similar state but, just in case, I repeated the procedure and found to my delight a perfectly preserved internal assembly. All that had happened was that the locking nut on the outside of the top plate had worked loose and allowed the two parts of the bi-metallic strip unit to rotate so that they were constantly in contact causing the short.

Five minutes with a screw-driver and a suitable sized spanner restored the unit to working condition. I resealed the unit with Araldite and after successfully testing it off the car, refitted it and now have a perfectly working warning light.

The accompanying diagram may help readers to identify the parts referred to in this article.

Cold Start Warning Light Sender Unit.

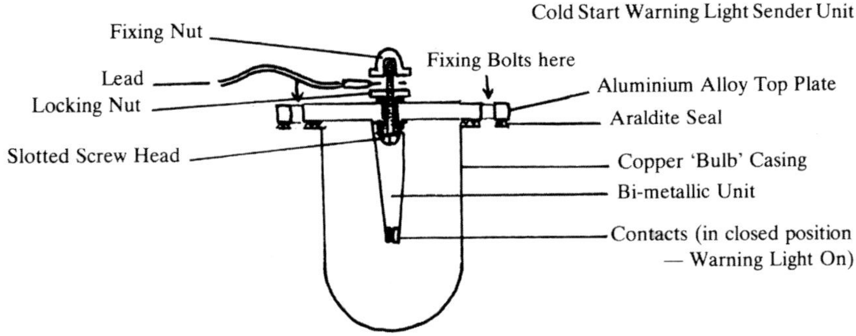

Chris Gough ***May 1985***

MUDTRAP IDLER

Photo 1

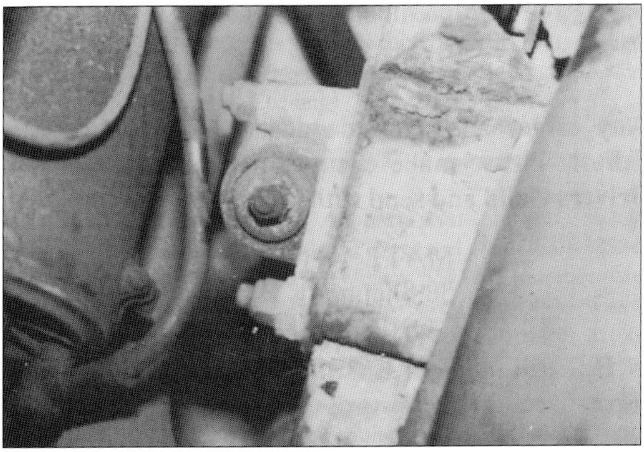

Photo 2

An excellent mudtrap on top of the chassis is the area between the tubular spacers adjacent to the steering idler.

Photo 1 shows a build up of mud to the top of the spacers, with Photo 2 showing what lurked underneath the mud – one large hole which would allow water to enter the chassis causing internal rusting.

Close observation of Photo 2 shows the remains of the sound deadening material which fills parts of the chassis, this acts very much like a sponge retaining water virtually indefinitely once wet.

The photos also show that the idler has been fitted with the modification to the filler plug, i.e. end float adjusting bolt and lock nut, but visual indications are that the idler has not received any lubrication in recent times.

The idler is not shown in most lubrication charts, as in original Rover literature the unit was classed as 'sealed for life', which roughly translated means the life of the oil remaining inside the idler.

Make a rule to top up the idler when topping up the kingpins, SAE 90 being a suitable oil. Remember firstly, before removing the filler plug, to clean around the top of the unit to prevent any road grit from entering. If your idler has the modification, it is advisable to readjust the end float if the filler plug has been removed for oiling purposes.

In this case, before refitting the filler plug, loosen the lock nut and run back the adjusting bolt a few turns, this will prevent the bolt contacting the shaft before the plug is fully tightened.

Replace the filler plug and tighten securely. Using a screwdriver adjust the bolt in a clockwise direction until it contacts the idler shaft, then back off fractionally, this will allow the shaft free movement but will restrict the end float to a minimum. The locknut should then be retightened. Do not be tempted to overtighten the adjusting bolt, this will result in stiff steering and will cause rapid wear to take place on the top idler bush.

On one idler which I recently came across, the lip of the top bush was completely worn away, allowing the collar to rotate against the alloy idler housing, a situation which if left would result in a worn housing. The modification obviously took care of excessive end float, but in this case actually hid a badly worn bush.

Colin Blowers ***May 1985***

CLUTCH AND BRAKE PEDAL LUBRICATION

Apropos stories to tell, some 5½ years ago, I experienced the brake and clutch pedals acting in concert on my '95'. All attempts at separating them failed due to the passage of time and possibly inability to get sufficient purchase.

As we were due to go on holiday I asked my local garage if they would have a go. On picking up the car later they had spent about 1 hour just attempting to lubricate the offending parts (spindle and sleeve) with a further suggestion that I take it back after 500 miles for further 'treatment'. As this would have been immediately after the holiday I decided not to.

Certainly the lubrication worked if only temporarily and some more positive action was required. I then decided to attempt to drill through the sleeve from the top and to put a piece of pipe in through which could be dribbled '3 in 1' oil. You will realise that to get through in this manner required drilling first through the floor and then the frame.

A small pilot hole, drilled to find position was not quite right but from that it was possible to put the next hole in the right place. The pipe I intended to use was ¼ inch O.D. nylon and my ¼ inch drill was barely long enough. However, I eventually succeeded in drilling through the sleeve, placing the tube in position and dribbling some oil down it. Lo and behold! I have had no further trouble with this problem and only need to dribble oil occasionally. The top of the tube is sealed off with a small plug to prevent ingress of dirt.

With no lubrication access provided I hope this may be of help to other members experiencing the same problem

Alan Sterne — ***July 1985***

CAM & ROCKER SHAFT

I read with great interest the article in last month's Overdrive on cam and rocker shaft removal, as I did this job myself a month previous on my 85,000 mile '95' and found badly worn rocker shafts. But I ignored the manual as far as drilling holes in the bulkhead to remove the rear shafts as both shafts can be removed from the front end with ease, with the aid of a Stan Johnstone Rapide removal tool. Which consists of a piece of tube slightly smaller than the shaft to be removed which will pass through the bearing surfaces and long enough to reach the shorter shaft once the longer front shaft has been removed. With an 'easy out' driven in or held in with an Allen screw in one end and a hole in the other for a (T) bar.

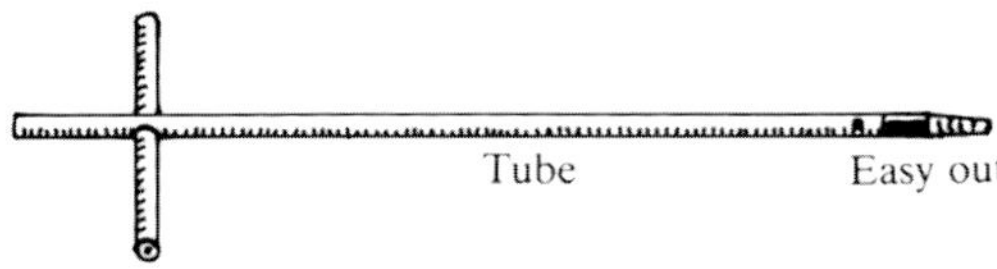

This job I did without too much difficulty on my own even though replacing rockers, followers, spacers and springs with one hand and pushing shaft back with the other, tested my back and my temper. But I must say that I think the manual leaves a lot to be desired on quite a few points during my fifteen years as a Rover owner.

R. J. Cundy ***July 1985***

GRAB-HANDLE RESTORATION

When I acquired my P4 the 'grab-handles' in the rear were very frayed and the metal tongue inside was showing through the material. Incidentally, these items, sometimes referred to as 'anklestraps' are in fact described as 'Pillar-Pulls' in the Parts Catalogue.

After removing them from the car, spring the chrome end to one side and you will find two more screws. They are usually tight, but once removed the strap can then be unfolded. If the tongue has split the material, the former can be stuck back with Evo-Stik to prevent the problem recurring.

Woolworths' Iron-on Tape can then be applied to the inside, after trimming any frayed edges with scissors. It is cheap, the right width and a good colour-match.

If necessary, clean the outside with Decosol, or spray upholstery-cleaner (which keeps the material drier). Chrome-cleaner will probably do wonders for the end-assembly, then all that remains is to reassemble and screw them back in the car.

Alan Gilham ***September 1985***

TIPS

I can offer the following method of replacing the spire clip slotted plates which have a maddening tendency to tear away from the hardboard base of the trim panels because the clips have rusted into the door itself.

The only things required are a 4 BA taper tap, obtainable from any good ironmonger and some 4 BA counter-sunk headed screws.

It will be found that the slotted plates have built-in hollow rivets, these will accept the 4 BA tap with no trouble, the only other thing is a small piece of tin plate – a bit of treacle-tin lid will serve very well. Drill it to correspond with the holes in the slotted plate, ease the rexine, or leather back on the edge of the door over the affected part, slip the tin plate in under the covering, insert screws, trim off any excess thread, re-glue the trim and you have a repair stronger than new.

F. L. Pearson — ***September 1985***

SERVICE YOUR WATERWORKS!

Now's the time to give a bit of thought to your P4's cooling system before the winter sets in (have we had summer yet I hear you ask?!!). In view of the excessive rainfall we've had in the last couple of months, I don't think your Local Water Authority would object to you using a few gallons in giving the waterworks a good flushing.

In actual fact, the Rover's system should ideally be flushed twice yearly, but if it is done religiously once a year without missing a year occasionally, this should suffice. I am making a point in saying this because having chatted to various members in the past they look at me aghast when I mention that I've just done my yearly flush. Some in fact, believe that anti-freeze lasts forever. The system must have at least 5 pints of anti-freeze in it at all times and should be replaced with new every year. Having a cast iron block and an aluminium cylinder head, this creates a 'dissimilar' situation which encourages corrosion and the inhibitors in the anti-freeze help to control this. Plain water, especially in an area with 'hard water' will rapidly corrode the waterways particularly the aluminium head, thermostat housing and water pump.

Thermostats do not go on for ever either and should be renewed every two years. Incidentally, it is imperative to fit an alcohol filled bellows type which always fails safe (open) as opposed to the waxstat type which fails closed. This of course can lead to excessive heat developing in the head and be a possible cause of warping and blown head gasket. It is never wise to run without a thermostat, as not only will you have a tepid heater in the winter, but the engine will not run very efficiently without it. Plus, an overcooled engine can tend to increase the petrol consumption slightly too.

Several days prior to flushing the system it pays to add a proprietary de-scaling compound to the cooling water and immediately after a run, open the drain taps on the radiator and block and allow to drain. If the flow seems slow, probe the taps with a length of wire. Once drained, release the bottom hose at the water pump connection pipe and place a garden hose in to the top of the radiator. Turn on the water from the tap and adjust the rate of flow to allow as much water as possible through the radiator and out through the bottom hose without making too much mess of your engine compartment.

Allow this to run for about fifteen minutes and then reverse flush the radiator by placing the hose in to the bottom hose and allow the water to run up the rad and out through the filler at the header tank. Again allow the water to run for about fifteen minutes. That should take care of the radiator, now you can concentrate on the engine.

Start by removing the top hose and then the thermostat and its housing. Ideally, it is better not to force any sludge through the heater box, so the heater should be cleaned separately from the engine block. The heater can be bi-passed by removing the two hoses from the heater and linked together with a short length of ½ inch copper pipe. This may seem a bit of a chore but it will leave the system as clean as possible and once done, it will be quicker to carry out on subsequent flushes. Wedge the garden hose in to the empty thermostat aperture and adjust water flow to allow the fastest circulation without spilling back out of the top. If possible, run the water for about half and hour, with a drain nearby, or course, and the bottom hose still disconnected.

This done, the heater can then be flushed by attaching the hose to the pipe at the top of the heater box and to save making a mess of your engine, attach a short length of hose to the bottom pipe and direct down on to the road. Run water for about ten minutes and then reverse the pipes and force the water up through heater until clean water is coming through the box.

Replace and tighten hoses having checked that they are in good condition. Leave the heater hose disconnected from the top of the heater. Check that the valve is open on the heater (if it is still operative!). Replace thermostat using a new gasket and with both the drain taps closed, slowly fill the system with water, adding the anti-freeze as you go. Fill until water starts to escape through the disconnected hose at the top of the heater. Connect and tighten hose and run the engine until hot water can be felt passing through the top hose, showing that the thermostat has opened. Check for leaks and with the engine still running undo the top hose on the heater box until a steady flow of water is running from it, with no air bubbles, and then replace and tighten the hose.

Take the car for a test run keeping a watchful eye on the temperature gauge. Check under the bonnet on your return just to see that you have no leaks. You will have to wait until the first winter's night before you are rewarded for your work with nice warm feet.

Stan Johnstone ***September 1985***

RENOVATION OF INTERIOR DOOR TRIM

First remove the inside moulding (2 s.t. screws in front top and rear channels or through front window surround in front doors) and carefully ease out the moulding. I would suggest covering the gap in the top of the door with a cloth to prevent the screws disappearing inside.

Remove the door and window regulator handles by prising back the cap over the coil spring and pushing out the retaining pins with a suitable punch, followed by the door pull. Remove the small s.t. screws under the moulding which hold the top of the door casing.

The casing is now retained by spire clips and two s.t. screws (front doors only) which retain the spring wire which keeps the top of the door pocket taut. Ease out the spire clips with a long broad bladed screwdriver placed close to each clip. If you are just removing the door casing to gain access to the interior of the door ease the casing away from the door whilst unscrewing the s.t. screws at the top of the pocket. This leaves the screws in the casing and does not release the spring wire.

Replacement is the reversal of the above. When replacing the moulding fix masking tape to the moulding adjoining the metal tags and mark the alignment of the hole on the tape. This enables the screws to be more easily located in their holes.

The cloth trim at the bottom of my doors was badly discoloured and rotten. Material of the correct colour was not available so it was necessary to purchase the lightest available shade of beige and proceed to a friendly dyeworks to obtain the required shade of green.

Remove the old trim from the hardboard and then carefully dismantle into individual items making careful note of how it was originally assembled, otherwise you will be in trouble. Sew in the new material as necessary which is easier said than done, taking care not to over-stretch – I used a domestic treadle machine.

Fix trim to the hardboard and re-assemble taking care to ensure that the eyes at each end of the spring will have the s.t. screws passed through them.

You should now have a smart door casing.

Derek Humphreys ***January 1986***

THOUGHTS ON LUBRICATION

The P4 being as robust and well engineered as can be will stand up to all kinds of abuse but its achilles heel, like all machines, is its inability to keep going without regular lubrication at all points. I hope the following views will be of use and of interest.

The Engine:

As I mentioned before, the engine oil should be changed every 3,000 miles. This is most important because beyond this point the oil deteriorates rapidly. You only have to rub the old oil between your fingers to notice how thin and 'dry' it becomes after a few thousand miles. The filter has also to be changed. I have heard people say they change the oil at 3,000 miles and the filter at 6,000 miles. i.e. change filter every other oil change. I personally don't think this is good practice as a clogged filter just gets by-passed and becomes useless allowing all the damaging silt to circulate round the engine. Good quality oil should be used as opposed to the cheaper varieties to be found in the supermarkets.

I like to think that the nearest modern equivalent to the oil originally specified, i.e. Castrol XL, is Castrol GTX. However any good multigrade oil is usable in the 20/50 range but I think the new 10/40 oils are a shade too 'thin' for P4's.

The Gearbox:

The gearbox oil should be changed every couple of years or at about 20,000 miles. The handbooks can be a bit misleading because at one time a 90 E.P. grade was recommended and then in later editions a 20 S.A.E. oil is stated. This came about because of a rather sluggish overdrive operation in very cold conditions using 90 E.P. So a 20 grade oil was them said to be better for all weather conditions. In actual fact both oils do the job but, of course, E.P. oils were really intended for the meshing of gears, so I think it is probably better than 20 S.A.E. This is particularly true if the seals on the gearbox are a little bit suspect.

The Rear Axle:

Some people seem to think the oil in the differential lasts forever. Whilst it does have a long life it does break down eventually. Therefore it should be changed every 20,000 miles or so. After a high mileage the oil becomes full of tiny particles of metal and if allowed to churn around between the gears, premature wear will ensue and lead to the dreaded 'whine from the back'.

Steering Box and Idler:

The steering box should be checked once a year and topped up with 140 (if you can still find it) or 90 E.P. If the seal is defective either replace it if possible or check and top up at regular intervals. The idler shaft needs topping with the same grade of oil just to the top of the bore of the shaft, no more or you will force the oil through the seal, when you replace the nut.

Swivel Pins:

I have of course, gone into detail before on swivel pins, but just to reiterate on what I said before, I think the ideal lubricant is a mixture of 20/50 and molyslip. This is slightly thinner than the 140 stipulated but it does encourage the oil to 'creep' more and therefore reach the parts other oils cannot reach! 90 E.P. oil will do the job just as good. These should be checked at least twice a year.

Front Wheel Hubs:

Again it must be stressed that lubricants do not last forever and ideally the front hubs should be removed every 20,000 miles and washed thoroughly in paraffin and replenished with new high-melting point grease. Incidentally,

LM on a tin of Castrol grease means high melting point. Never did quite understand that! Do not over-tighten the hub bearings as this will only result in excessive heat and cook the grease. Always back off one hole when the nut has just bitten onto the bearing.

The Propshaft:

Don't forget the poor old propshaft. An occasional grease should be applied to the sliding section of the shaft and on earlier models to the universal joints as well.

All these lubricating jobs can be a chore I know, but they do keep the wheels turning and your Rover will go on indefinitely with a regular servicing routine. Incidentally 'Hermetite' are now producing a straight 20 oil in handy bottles to top up the damper or dampers on your S.U. carburettors. This is much better than 20/50 which is really too heavy or 3 in 1 which is too thin.

I keep a small notebook in my car and keep a check on the various lubricating jobs otherwise it is surprising how the miles build up and you can never remember when you last checked your various P4 duties.

Stan Johnstone ***January 1986***

BAKELITE—THE CURE FOR HOT S.U.'s

Reference to your hot starting problems mentioned in your editorial in Overdrive 45.

I experienced the same problem on my 1951 Cyclops and found the reason is vapour locks building up in the carburettors (not in the Cyclop's engine-bay-mounted fuel pump as might have been expected). Obviously the carbs got too hot after several minutes of rest (without the cooling air stream of the radiator fan) and the symptoms were exactly as you describe them.

The cure was to isolate the carburettors from the cylinder head: bakelite plates were cut to the same shape as the flange gaskets and were fitted together with two gaskets between carb. body and cylinder head. On my 75 the bolts are just long enough to accept the additional thickness of the bakelite (about 4mm thick), Thin isolating washers were fitted under the nuts to prevent heat flowing to the carbs via the bolts and nuts.

This treatment completely cured the troubles several years ago. When you look into the engine bays of modern cars, in most cases you will find isolating pieces between carburettor and head, so the problem seems to be quite common.

Michael Mayr-Harting, Austria *January 1986*

HOW'S YOUR AUNTIE?

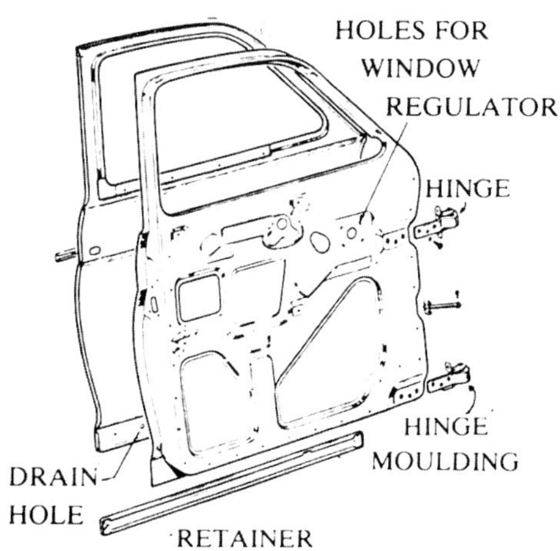

1. The drainage of water from within the doors is intended to be by way of several drainage holes behind the stainless steel mouldings which run along the bottom of each door. These mouldings are held to the doors by 4 or 5 retainers, the nuts on which are accessible along the inner side of the doors when in the open position.

At Auntie's age blockages can occur so that water may remain inside the doors for quite a long time. A clear-out is worthy of recommendation.

To aid removal of the mouldings it is worthwhile to carefully wirebrush the exposed ¼" dia. threads and nuts, and soak with penetrating oil before attempting to undo the retainer fastenings. It is probable that 1 in 4 of the retainer threads will shear off, even when care is taken to progress carefully. It is not essential that every retainer is replaced on completion of the work but the margin for losses is not so great that some spare retainers should not be considered.

When the moulding is removed from a door the drainage holes in the door can be cleaned out. The moulding can also be cleaned out after each seized retainer is given a light tap and slid to the moulding end where the slotted moulding form enlarges to allow removal. The retainers can be scraped of scale, and the threads wirebrushed and waxoyl applied generously.

With all surfaces cleaned the moulding(s) can be replaced and the reward found to be that rainwater will run out at the ends of the moulding as originally intended.

2. The weight of the wind-up window glass in a door is considerable. It is *not* supported directly beneath the glass but by the 4 screw attachment to the door's inner panel, but so to speak, at arm's length. In the case of aluminium R.H. front (driver's) doors – most models, the holes in the inner panel can become considerably elongated by the pressure exerted by the setscrews which have not remained tightly fastened.

WINDOW REGULATOR (WINDER)

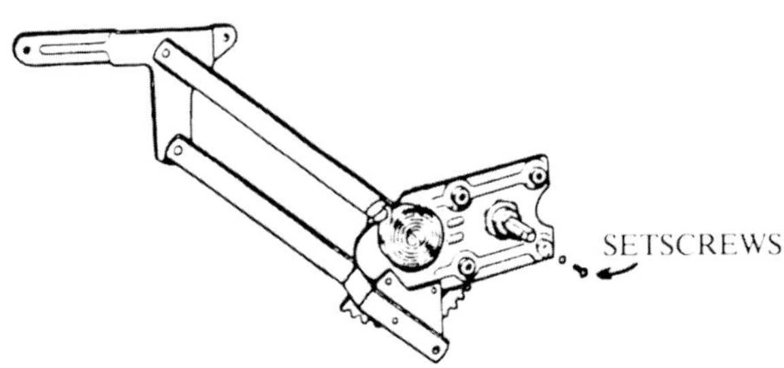

The window, receiving less support, will have started to lean against the rearmost support channel. If the window whilst not fully wound-up is lifted to its normal alignment and the aforementioned screws tightened all will be well again but only for a short period. A longer lasting repair can be made if the owner is prepared to make a mild steel plate which will assist the door panel to support the window regulator.

STEEL PLATE USED TO IMPROVE WINDER MOUNTING

(Shown in position in door recess)

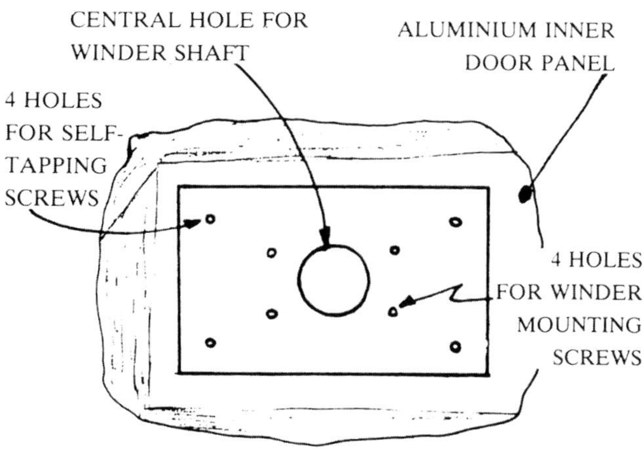

The plate is of 18 to 16 gauge in thickness, having a central hole of about the same size as already in the door inner panel which accepts the winder shaft. The plate is then drilled with 4 holes to accept the setscrews which mount the winder. Paint the plate to prevent corrosive interaction between the steel and aluminium. When the paint has dried, mount the winder loosely with the new plate as an inner door panel reinforcement. The mounting screws might best be used without their shake-proof washers but lightly coated on the threads with paint, mild 'Locktite' or similar. The extra thread length should compensate for the steel plate thickness. Ensure the window is properly aligned and tighten the mounting screws. The window when fully wound up should, for instance, not trap the quarter light glass.

Drill the 4 holes shown near the steel plate corners to accept 4 self tapping screws which secure the steel plate to the door inner panel. The mounting should now ensure firmer window winder support and not allow elongation of the holes in the steel plate.

On my own car the window guide channels from within the door were removed for repair. They were straightened and generally reformed carefully before replacement.

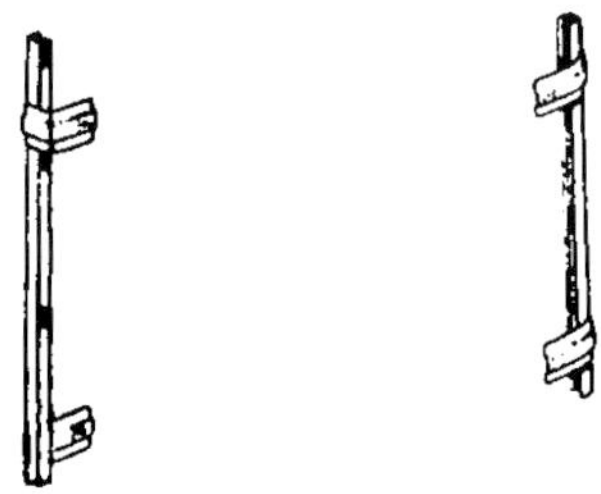

The window of the driver's door which had always been difficult due to frequent usage throughout the car's life now operated smoothly and hopefully the winder gears can expect a longer life.

Although not mentioned in the foregoing work description, the door trim panel is not completely straightforward to remove entirely. The problem is that after most necessary items have been removed or released (door interior handles, trim clips etc) the leather cloth still passes beneath the wooden door casing and is further retained at several points beneath the casing. Often work can be done with the door open and the panel lifted to an angle from the door. Sometimes a previous owner will have ripped the leather cloth locally at the top retention points – this can generally be disguised by the overlap of the wooden casing.

3. So after all that – if the bottom door hinge has allowed door droop but still has some life left before need of replacement, the opportunity to re-set the door by adjustment of the hinge attachment to the door is presented at this time. A jack with a broad head will be useful to support the open door whilst adjustment is made, and an open ended spanner probably the best to slacken the setscrews and retighten them after the door has been lifted slightly by use of the jack.

The actual condition of the bottom hinge is probably poor if water regularly enters its housing area. It can be expected that the hinge pin is not a pivot as designed but has rusted firmly in its hole, and that the pin turns, without restraint, in the hinge frame upon door operation.

But these are just minor things! "Sorry Auntie".

May 1986

DWELL ANGLES FOR P4s

Whatever next I hear you ask, dwell angles for P4's, far too modern an idea! Well in fact during the production of the P4, dwell angles for the car's ignition setting were never published but with more sophisticated equipment becoming available to the DIY mechanic they can now become useful. All right you may ask, we have managed so far without this information, so why bother now? This is very true for most of us but I do know that there are some enthusiasts among us who like to get things spot on.

I have always considered myself a dab hand with a set of feeler gauges, but after using a dwell angle meter, I have been surprised how far out you can be with the points gap setting. After renewing or adjusting contact points the timing should always be checked by the way.

Of course the use of a dwell angle meter will probably make very little significant difference to the performance of the car but over a long mileage the petrol consumption will be improved. Another useful aspect of the use of a meter is that it also shows up any wear in the distributor: i.e a reading fluctuating outside the ± 3 or 4 degrees tolerance allowed will indicate wear either in the cam lobes of the distributor and/or wear in the bush. With this situation a satisfactory running condition will not be obtained.

At this point, I feel I should state exactly what the term dwell angle means. It is the number of degrees of distributor cam rotation for which the contact breaker points remain closed during one ignition cycle. This is also important because the size of angle determines the length of time available for the primary current to build up in the coil and hence the size of the next spark.

All very technical I know, but I thought that with tuning equipment so readily and cheaply available that some of you might find the following figures useful.

All six cylinder models except '110'

The dwell angle is 38° ± 4°
The '110' is 35° ± 3°
The '60' is 60° ± 3°
The '80' is 60° ± 3°

For those of you that are quite happy with your engines as they are, do not despair, your engines are probably within the limits with your conventional settings anyway, but for those perfectionists among us I hope the above figures will be useful.

Stan Johnstone ***July 1986***

PEDAL SHAFT LUBRICATOR

To overcome the problem of stiff brake and clutch pedal fulcrums and tandem operation, as no sensible method of oiling could be found, I embarked on the following modification which has proved very successful. I believe many P4 owners would benefit from curing this common ailment.

Stripping Procedure

1. Referring to Fig. 1 — Withdraw brake pedal lever (5) from spindle (12) after firstly removing pedal rod (15), clevis pin (23) and fixing bolt (13).
2. Referring to Fig. 2 — Remove clutch cross shaft balance spring (where fitted) and clutch return spring (26)
3. Remove two bolts securing spherical bearing housing to gearbox (31).
4. Remove two pins (13) securing pedal spindle to connecting tube and connecting tube to cross shaft and lower cross shaft (12).
5. Remove clutch pedal rod (7) and bolt (4) securing clutch pedal lever to spindle.
6. Withdraw connecting tube (16), clutch pedal lever (2) and thrust washers (3).
7. Remove pedal spindle from housing and clean. Remove any rust with fine emery cloth.

Drilling Pedal Spindle
The use of a lathe is required for this exercise. Any garage or engineering firm with turning facilities can easily carry out this task.

PEDAL SPINDLE

Hole 'A' — Centre drill in lathe then drill ⅛" diameter to a depth equal to ¾" beyond housing bearing position. (Fig. 3).
Hole 'B' — Drill ⅛" diameter in brake pedal fixing bolt groove intersecting with Hole 'A'.
Hole 'C' — Drill ⅛" diameter ½" beyond bearing position (Fig. 3) intersecting with Hole 'A'.
Carefully remove any burrs with fine file and polish spindle with fine emery cloth.
Check fit in housing.

FITTING OIL NIPPLE

Fitting Oil Nipple (Refer to Fig. 4)
Take brake pedal lever and using centre punch pop mark the middle of space between lever for pedal rod and master cylinder operating lever in alignment with centre of hole for lever fixing bolt. Drill 13/64" diameter and thread hole ¼" BSF using tap and tap wrench. Carefully deburr hole in bore of pedal bush. Fit standard ¼ BSF straight grease nipple and check fit on spindle. Ensure blanking plug is secure by giving smart tap in centre with light hammer.

Reassembly
Clean all components.
If oil seals in housing are badly worn now is the time to replace (9 and 10 Fig. 1).

Reassembly — reverse of stripping procedure.
With a grease gun containing EP90 or similar oil, inject oil via nipple into housing, by careful feel keep injecting oil until resistance is felt – when air is expelled and housing full of oil.

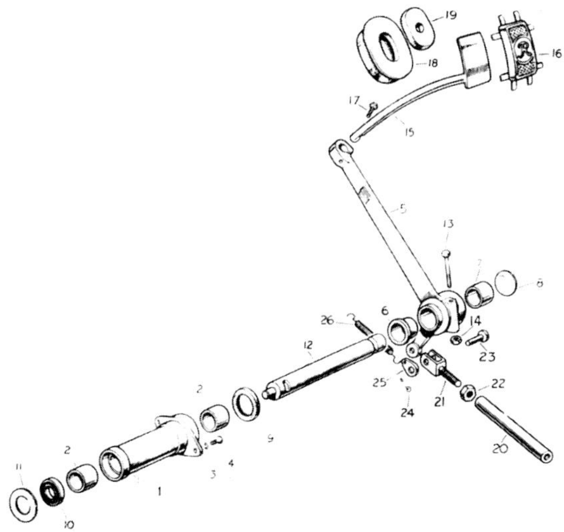

FIGURE 1

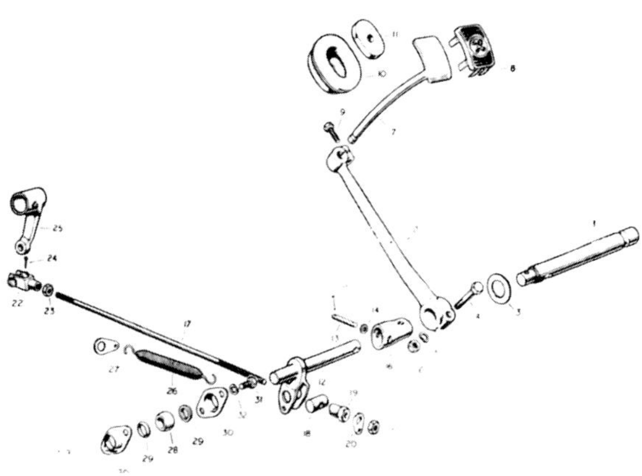

FIGURE 2

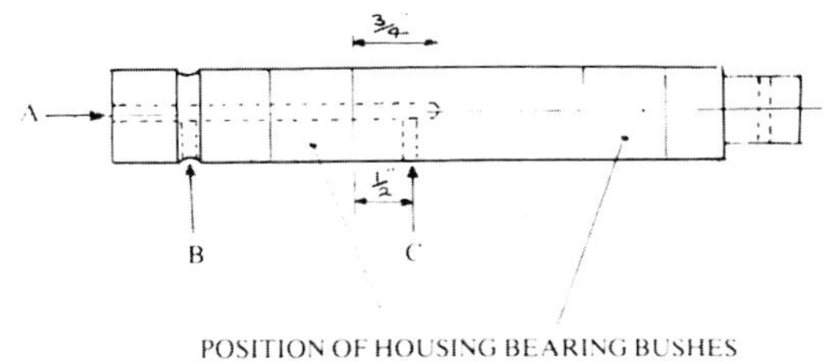

FIGURE 3

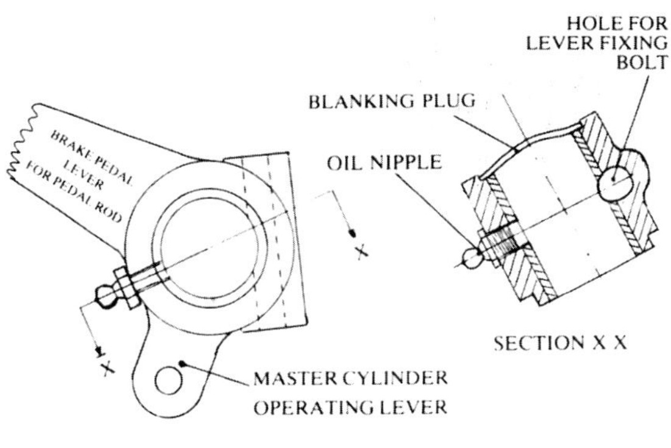

FIGURE 4

Conclusion

You will now have smooth pedal operation transmitting the correct feel to your boot as when new. By simple, quick attention at six monthly intervals to make up any loss of oil precise pedal operation will be maintained.

Andy Doe *November 1986*

GET OUT AND GET UNDER

I, too, have been painting the underside of my P4 and heartily endorse the "little bit at a time and stick to it" advice you gave in your November editorial.

At first I found the sight of my '100' so daunting from below, as it stretched away into the distance, that I very nearly didn't start at all. But little by little, an hour or so at a time, transistor radio at my side, I have wirebrushed and scraped, Jenolited and Smoothrited, until – I have to admit – things are looking really quite presentable down there. I now have a pleasing speckled appearance and the family are expert at holding conversations with my feet. But to my surprise I've enjoyed it, at least until the chill striking up from the concrete garage floor, even through a couple of layers of old carpet, becomes too much. Then I retire indoors to commune with the teapot and restore my circulation. Until the next small step.

One of the most useful aids to this multi-stage marathon has been a household radiator paintbrush – the type with a long wire handle. This you can bend in any number of permutations to get at the host of "important little places" that lie in wait to challenge the P4 owner, especially round the battery box and rear outriggers. Combined with a couple of artists' brushes, I've found I can easily reach the parts other brushes, etc.

I suspect you'll be reporting the completion of your "paint job" long before I'm half way finished, but I'm determined to make it..... "a little bit at a time".

Dennis Jones ***January 1989***

FUEL PUMPS

I have recently had to have new contacts fitted to the main SU petrol pump in my Rover '100'.

In order to keep the car on the road I wished to remove the main pump for servicing, but to retain the reserve pump in use, which meant blanking off the pipes to the main pump.

I therefore made the plugs (2) which are detailed on the enclosed sketch. As it turned out to be a job requiring certain vital dimensions, I thought it might be of interest to '100' owners to know the details.

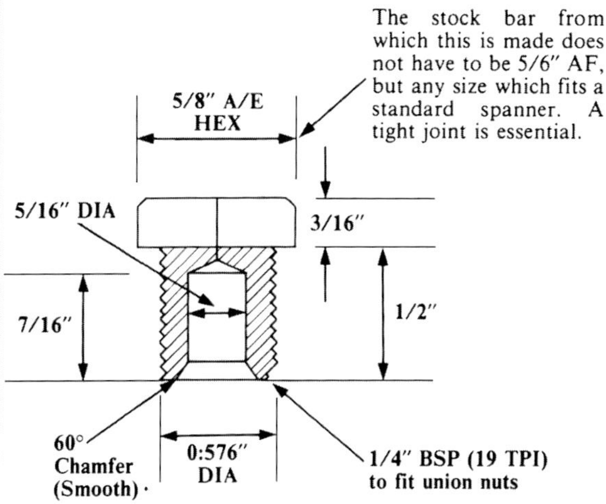

ROVER '100'
(TWIN PLUGS)

PLUG TO CLOSE PIPE CONNECTIONS TO SU PETROL PUMP WHEN USING ONLY ONE PUMP WHILE OTHER IS SERVICED.

K. Stocker — *March 1989*

ROVER OVERDRIVE ELECTRICAL CIRCUIT

The Rover '80' and '100' overdrive system incorporates some rather unusual and elegant electrical circuitry which merits explanation. The unusual feature is the way in which the overdrive relay coil is de-energised when the overdrive is disengaged. Instead of the relay coil being de-energised by switching off its current, which is the normal method, the relay coil is de-energised initially by applying battery voltage (-12v) to each end of its coil.

The complete electrical operating sequence of the overdrive is as follows:

Overdrive disengaged

The overdrive switch on the steering column is OFF; so there is no current path from the overdrive fuse, through the overdrive relay coil or solenoid coil, to the bottom earth line; hence, both the relay and solenoid remain de-energised. The ball valve in the overdrive hydraulic system, which is operated by the overdrive solenoid, remains seated. Refer to page C18 of the overdrive manual for an explanation of the overdrive hydraulic system.

Select Overdrive Engaged

Before the overdrive can be engaged, the gearbox switch must be closed by having top gear selected; and both throttle switches must be closed.

The throttle switch with the vertical plunger is set so that it is *closed* when the throttle is between 0 and one half open. The throttle switch with the horizontal plunger is set so that it is *open* when the throttle is between 0 and one eighth open. Therefore, for both switches to be closed, the throttle must be between one eighth and one half open.

Having engaged top gear and set the throttle so that both switches are closed, overdrive is engaged by moving the steering column switch down to ON. This makes the circuit from the overdrive fuse, through the relay coil and relay resistance, to the horizontal earth line at the bottom, thereby energising the relay.

When the relay is energised, the RL contact closes so that current flows through the overdrive solenoid coil, thereby energising the solenoid.

When the solenoid is energised, it lifts the ball valve off its seat to direct high pressure oil to the overdrive operating pistons.

Kickdown

When the kick-down switch beneath the accelerator is operated by pressing the accelerator hard down, the KICK TO CLOSE switch closes. This applies battery voltage to both ends of the overdrive relay coil, thereby neutralising (de-energising) the coil and causing the RL contact to open. This breaks the feed to the solenoid coil and de-energises the solenoid. The ball valve reseats to shut off the oil supply to the overdrive operating pistons, and allows the oil from the pistons to pass to exhaust; the overdrive then moves into disengagement.

On releasing the kick-down switch, the system reverts to its previous state.

Select Overdrive Disengaged

When the steering column switch is moved to off, battery voltage is applied to the lower end of the relay coil, via the horizontal plunger throttle switch (providing the throttle is more than one eighth open) and the OFF terminal on the selector switch. As battery voltage is also being applied to the top of the relay coil, the coil is neutralised and therefore de-energised, which in turn de-energises the solenoid by the opening of the RL contact.

Once the RL contact opens, the circuit between the overdrive fuse and the bottom earth line is broken. The de-energising of the solenoid reseats the ball valve, so that the hydraulic oil is released as previously described and the overdrive moves into disengagement.

Mike Austin ***March 1989***

COOLING DOWN YOUR P4

I have been asked many times for advice on the problem of overheating on the P4. I have also read articles in the Overdrive of members who have gone to great lengths to solve the problem, one member ended up fitting an expansion tank return system. Others run around without the thermostat.

Let's get one thing straight, apart from the 1950 to 1953 cars, no P4 should overheat, if it does then something is wrong, let's face it, it is, compared with modern day cars, a low pressure water system, it has approx. two gallons of water, and two gallons of oil. I mention oil, because apart from its lubricating properties it is also used as a coolant.

Let's look at what could be causing the overheating.

Thermostat – always worth replacing.

Blocked or partially blocked radiator.

Ignition timing and valve timing.

Carburettor settings.

Head gasket leaking.

Brakes binding making the engine labour.

Slipping fan belt.

Water pump.

Assuming that the average P4 owner being a type of person that keeps his car in good service order, has had all the above checked, is still scratching his head, the following could well help in solving the problem.

First drain the system completely, i.e. both drain taps, flush through fresh water via a hose from a water supply, reverse flow the flushing by putting the end of the hose pipe over the block tap and then the radiator tap, yes, I'm sorry this will probably make a mess of your polished engine compartment, but never mind, after the flushing and there is no more sediment or brown coloured water coming out let the system completely drain down. Close the two taps, make up your mixture of good quality anti-freeze of at least 33⅓% but these days it is recommended a mixture of 50% is correct, remember, anti-freeze has properties built into it to protect engines from corrosion and overheating, on modern vehicles anti-freeze is left in all year round and is mixed at 50%, there is no reason for not giving your P4 the same protection, especially with our aluminium heads etc. It must of course be checked from time to time for strength. Fill via the radiator cap slowly allowing the mixture to circulate and settle, keep topping up, have the heater controls on the dash set to hot, when you think the system is full, put your hand down and compress the bottom hose, do this several times, working it like a pump, you should see air bubbles and movement in the top of the radiator, when the air bubbles have stopped, check and top up the radiator again, never fill to the neck of the filler, always allow at least ½" to ¾" below the neck, fit the radiator cap and start the engine, allow to reach normal temperature, check for leaks and allow to idle, switch off after 15 minutes, allow the system to cool down, check and top up if required.

The last point, and I am not having a dig at the Rally and show members who only take their P4's out in the summer, but I can't help noticing the numbers of cars with the heat control left in the off position, air locks do occur in engines from time to time, and leaving the control in the off position does no good, firstly by having the control in the hot position allows the cooling water to circulate through the heater matrix and helps to keep it clean, and secondly, by allowing the coolant to properly circulate throughout the engine. I have often cured a member's problem of overheating by opening the control to the hot position and taking the car for a run, take a container of coolant with you, because you may find your temperature gauge doing funny things, if it does, stop, allow the engine to cool down, and then top up the radiator as described. You don't of course have to run around all summer with the heater on, you can of course return it to the off position, or have it in the hot position with the main control knob in the fresh air position, this allows water to pass through the matrix, but not into the car.

If as I do, you tow a caravan, you will find after a long drive in hot conditions, or climbing long hills, i.e. Dorset, Cornwall etc., the temperature gauge will increase to the middle of the white position, but should not go into the red.

Ian Jones ***September 1989***

NOTES ON FITTING A RADIO CASSETTE TO A P4 '110'

I can think of no very good reason why a car like the P4 should not be up-dated with the latest gadgets in modern technology as long as it is done with discretion and taste. I have fitted to my own P4, radials, quartz lamps and electronic ignition and it is none the worse for it and I certainly feel just that little bit safer as a result. I thought a good quality radio would be a worthwhile addition and so this is a blow-by-blow account of what is entailed.

All the radios you can buy now are negative earth so you will have to reverse the polarity of the car before you switch on the set. Take out the battery and put on one side for the moment. Drop the battery leads below the car and remove the lugs. These must now be fitted to the opposite cables. Gettit? My P4 had integral lugs and there was no way of saving them so they had to be sacrificed (you can melt them off with a heavy-duty soldering iron). I had to fit another neg. lead because the original one was too short (you have to turn the battery round 180° which means the neg. lead will be too short to reach the neg. post on the battery. Are you still with me? NOW and this is MOST IMPORTANT – TAPE UP THE ARMOURED CONDUIT ON THE LEAD WHICH WHEN FINISHED WILL BE GOING TO THE POS. LUG. THIS WILL PREVENT THE CONDUIT CREEPING ALONG THE CABLE AND CREATING A SHORT-CIRCUIT. So much for the battery.

Now for the clock.

My clock has two wires going to it which means that the '110' owner does not need to alter the wiring. Owners of lesser vehicles which have only one wire feeding the clock (and here I refer only to power supply NOT the wire feeding the bulb) will have to remove the clock, insulate it completely and reconnect it. TAKE EXPERT LOCAL ADVICE ON THIS MATTER!

Now the dynamo.

Re-polarise the dynamo by letting the engine idle and close the cut-out points. If you then open the throttle slightly you should see a discharge on the ammeter. Let the instrument panel down and change over the connections on the ammeter. And that's the dynamo done!

Now for the aerial.

USE THE PROPER ONE – nothing looks worse than the wrong aerial on a car like the Rover. (At first I fitted a stick-on aerial to the inside of the windscreen – it looked neat and unobtrusive but it didn't work!) Eventually I fitted one I had had on a P6 and it looks a treat! Take off the visors, the mirror, the narrow bit of head-lining and the windscreen moulding. You will then see a hole of about 4 inches in diameter. Find the centre of this and drill a hole through the roof to match your aerial mounting. Fit the aerial remembering to put some sealing compound between it and the roof and lead the wire down the side of the screen to the back of the dash. DON'T FORGET TO SCRAPE THE PAINT OFF WHERE YOU MAKE YOUR EARTH CONNECTIONS. I had to drill another hole at the top corner of the screen to get the wire tucked away neatly. This done, head-lining, visors, moulding and mirror can be put back.

Now for the speakers.

Because all sets nowadays are stereophonic you have two speakers which in my case were square because I thought they would be easier to fit. I put these under the front seats out of the way, each one in a hardwood box which, according to the experts in 'in-car entertainment' has the effect of throwing the sound forward. I suspended them from the frame of the seat by 'L' shaped pieces of mild steel. Drill a hole in the bottom of each box and take the leads under the carpet up to the set.

Now for the receiver.

Take off the middle cupboard door (which is NOT solid mahogany as the Road Test Report would have you believe) and substitute a veneered piece of ply (or even solid mahogany!). The cardboard pocket and second bit of card will have to come out (you need all the depth you can get) and be discarded. Make two pieces of angle-iron and pop-rivet them vertically to the dashboard frame now exposed. The radio fascia board will be screwed to this using self-tapping screws which, unhappily, are not brass finished. Cut a hole in this board and feed the set through a line-fuse from the vacant terminal on the ignition switch. The set should be earthed to the chassis of the car.

Remember, you get what you pay for, as in everything. I bought Radiomobile in the fond, perhaps vain hope that it was not Japanese owned.

Suppressing the dynamo, coil, etc. will have to be done on an empirical basis.

R Jessop ***November 1989***

P4 DOOR HINGES

As members are aware door hinges eventually wear and the pins can even fracture resulting in sagging and rattling doors. Repair of the hinges, when removed, is not difficult provided one has and can use the necessary equipment – it is their removal when bolts have rusted and the bottom of the front wing is in a fragile state which causes problems. However it is a job worth doing so proceed as follows.

Remove the interior door handles by pressing out the retaining pins whilst holding back the spring loaded escutcheons. Remove door mouldings – the front mouldings have four S.T. screws in the glass run channel and two in the front face of the door and the rear doors have all six screws in the glass run channel. The door casing is then removed, this being retained at the top by S.T. screws under the moulding, 2 S.T. screws at the top of the door pockets on the front doors and numerous spire clips. Do not remove the screws at the top of the pocket from the casing otherwise you will lose the ends of the expanding curtain wires which control the opening of the door pocket. When prising out spire clips, make sure the screwdriver or tyre lever actually contacts the clip before exerting pressure as this reduces the possibility of the clip retainer pulling out of the hardboard. Remove the pin from the check strap and then the bolts securing the hinges to the door noting the position of any shims.

Front Hinges

Remove the A post rubber retainer – S.T. screws, and the set bolts securing the rear of the wing to the A post and the bottom of the wing below the wheelarch moulding thus enabling the bottom of the wing to be pulled outwards taking care at the same time to release the rubber seal between wing and splash panel which will be stuck to the wing with underseal. Remove the scuttle trim panel from inside the car again retained by spire clips. The Workshop Manual says pull back the end of the sill mat to gain access to the retaining screw underneath but I find it preferable to remove the mat. It is now possible to gain access to the hinge fixings – upper hinge, two screws in the face of the A post and a nut accessible behind the scuttle panel – lower hinge, one screw in the face of the A post, one bolt through the sill with the nut usually well rusted and a nut accessible behind the wing.

Rear Hinges

Remove sill mat and stainless steel wheelarch moulding which is secured by a nut inside the wheelarch and a S.T. screw through the leading edge. The upper hinges are secured by two screws in the door post and two screws accessible through the hinge aperture. The lower hinges are secured by one screw in the face of the door post, a bolt through the sill and a nut behind the wheelarch moulding.

Repairing the hinges involves grinding off the welds securing the locating washers on each end of the pin, driving out the old pin, drilling and reaming to fit a suitable oversize pin and rewelding the locating washers to the new pin. Quite often the section of the hinge which bolts to the sill is rusted and requires repair by welding. Also make sure that the oil groove is clear to the pin.

Replacement is the reversal of the above but includes the additional work of correctly aligning doors and ensuring that the rubber seals inside the front wings are correctly positioned and sealed with underseal to prevent ingress of water.

You could well find on dismantling that further repairs are required to the sills in the vicinity of the front hinge boxes and also the wings behind the wheelarch mouldings.

I would be happy to repair hinges for members at very reasonable rates if they were forwarded to me.

Derek Humphreys ***January 1990***

WINDOW FRAMES AND CHANNELLING

It's worth doing both jobs at once as to do just one means removal of the other and both are tricky to remove and replace. All the screws in the channelling are difficult to get out, particularly those at the side rather than at the top as these have more moisture trickling past them over the years. A really sharp, square screwdriver is needed and perhaps a pre-soaking in WD40. Even then you will probably have to mutilate one or two screws but you may be lucky. Cherish the screws that you manage to remove as they are scarce. When you have all screws removed the wooden window frame and then the channel should come out. In my case the wood joints were all loose and with the moulding on the bench it was an easy task to prise the components apart keeping all the plywood fillets that join the corners. It's worth marking them or keeping them in the same position to aid re-assembly.

Clean all the old glue from the joint faces, the fillets and their grooves. Then re-assemble all the joints with woodglue and either clamp or bind the structure together until the glue is cured.

Obviously some care must be employed here to ensure that the structure is not twisted or otherwise misaligned. Once it is a complete and solid frame it is much easier to clean off the old varnish using a craft knife blade as a scraper and to smooth off with progressively finer sandpaper or wet and dry used dry. When you are satisfied with the surface clean off all dust with methylated spirits and allow to dry. I then used polyurethane varnish in about four coats, rubbed down between coats and finished off with a light going over of T-Cut or even Solvol to smooth the surface. Lastly put a bit of wax polish on the wood to protect further. It is important to varnish the inside edge that is visible from outside of the car when fitted, that is, the bit closest to the glass at the bottom. I of course only noticed this once the first window had been re-assembled. Originally the mouldings had a similar material to headlining wrapped around the inner section – presumably to alleviate rattling or draughts. This went under the metal tabs that attach the moulding to the door. I replaced this with black felt glued in place. The tabs are of course cleaned and painted and replaced. So now you should have the mouldings complete.

Now to the door and channelling. You should already have the channel out and this will give you access to clean up any corrosion present on the aluminium (or steel) – removing draught strips, cleaning and painting as you go. The channelling that is currently available from specialist automotive suppliers has no chrome strips; it is black all over but is very easy to handle, cut and bend, and produces a lovely smooth operation that is entirely draught free. Its construction is an improvement on the old type as the illustrations show. It would be necessary to drill through the back of the old type in order to fit the screws. With the new ladder type it is only necessary to choose a slot that is closest to the existing holes in the door and push the screw through. The whole 'ladder' is covered in a malleable rubber material rather similar to Dum-Dum that stretches easily for sharp bends. It is necessary either to use the old channel as a pattern or measure accurately so that the ends of the channelling fit snugly into the vertical glass runners. Cutting is easy, simply flatten the channel at the relevant point as in Fig. 2 and with snips or heavy scissors or pliers snip at points A and as the rest is either felt or rubber, scissors will easily cut through. It can then be bent back to shape without damage. Achieving the curves for the door aperture is a matter of trial and error and it is worth spending a bit of time to get it right. The flexible channel is held in place by two screws only but the moulding has six screws on the rear doors that attach it. Having affixed the channel it is worth, using a wax pencil, marking the outer surface of the wood to indicate the positions of the metal tabs. Offer up the moulding to the door and carefully push the tabs between the channelling and door. Using a narrow but sharp bradawl or other tool line up the holes in the tabs with those on the door and the channel utilising a suitably placed slot. It is inevitable that some will not line up but the forgiving construction of the channel will allow a screw to be inserted at an angle and still do the job without fouling the glass. The bradawl is important as an aid to line up the position for the screw. Use a nice sharp screwdriver of the correct size and the job will be easier. I bought about 180 inches of the channel which is more than you need but will allow for a mistake in measuring, and this cost less than twenty pounds.

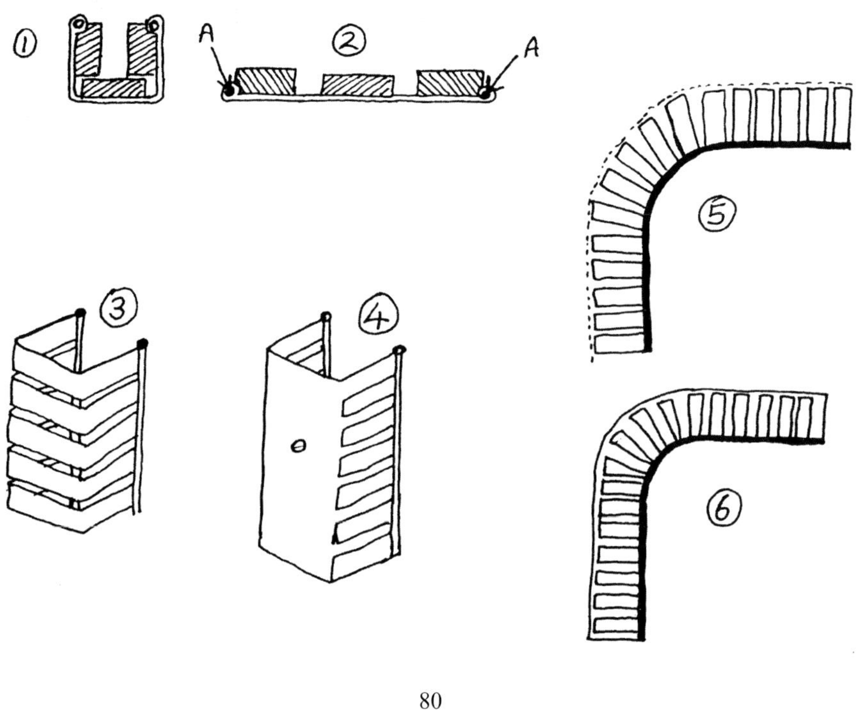

Key to Illustrations

1) Cross section of the flexible window channel
2) Flattened channel ready for cutting at points A
3) Skeletal view of new type channel – screws are inserted through suitable slot
4) Old type channel with solid back that would require drilling
5) New type and 6) old type channel at fairly acute bend, i.e. forward edge of rear door. The stretchy rubber on 5) covers the ladder even at the apex of the curve. Old type channel would need care to bend accurately

Have a go and say goodbye to rattly, draughty old windows for another thirty years!

Matt White *November 1990*

P6 SU CONVERSION FOR THE '80'

Much interest was shown in Pat Moores P6 SU conversion for the '80' so he has put together the following notes to help you carry it out.

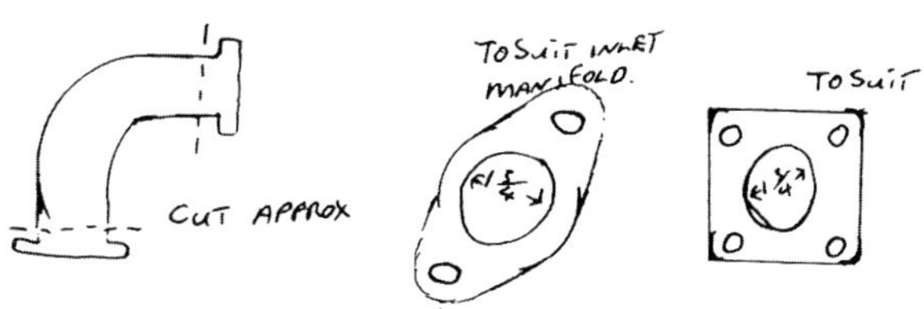

Firstly the manifold adaptor – I used a 2" plumbers elbow, which I cut down as small as possible in order to get the carburettor close to the manifold. The adaptor plates were made from 5⁄16" thick mild steel plate, gaskets were used as templates to make these.

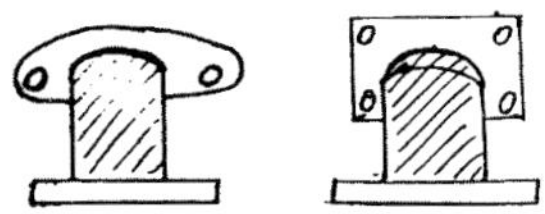

When complete, the adaptor plates were brazed to the elbow at the appropriate positions.

The air filter was adapted using an SU Carb pipe and a piece of 3" tube and original parts.

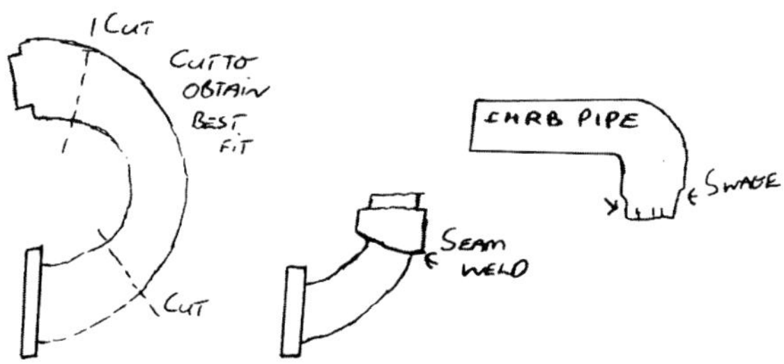

Throttle Linkage

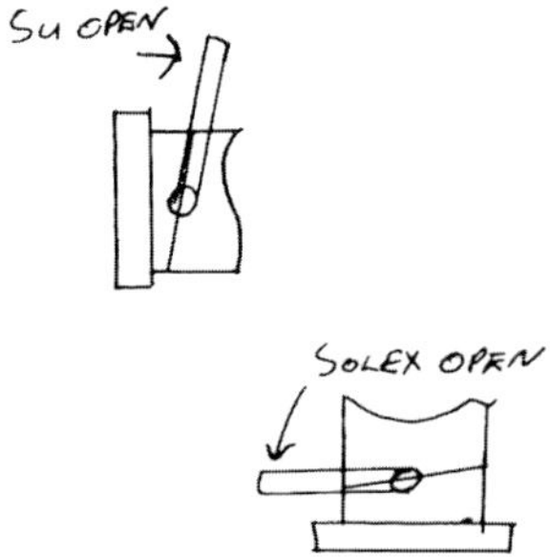

The flexible drive on the linkage must be replaced by a solid piece of bar because the SU throttle opens the opposite way to the Solex; causing the cable to unwind on the flexible drive.

The pedal linkage should be altered on the bulkhead to suit.

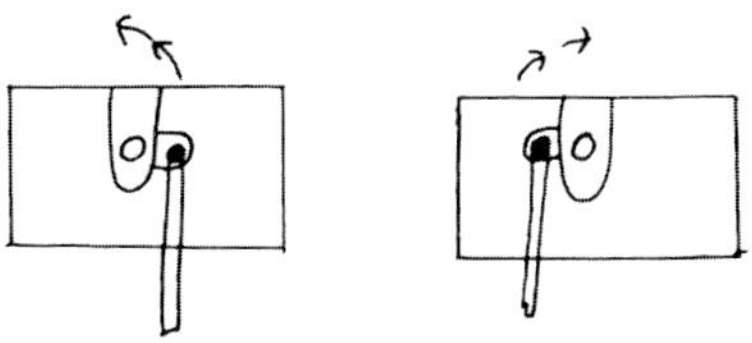

The linkage is all trial and error to obtain the best setting. An additional spring was also fitted to assist the throttle return.

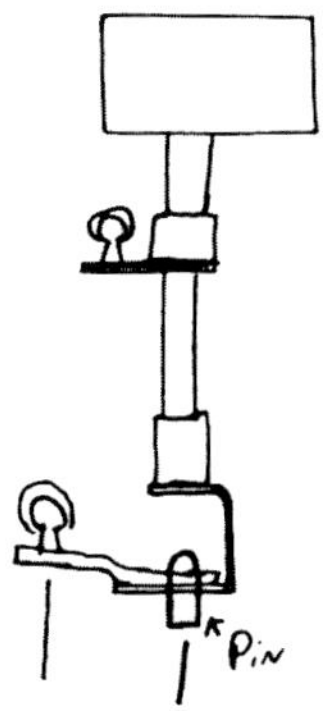

A bearing arm for the throttle linkage was made, using a piece of steel and a bronze bush.

The bearing arm was fitted to the car using one of the manifold studs.

Two levers were made keeping the hole centres equal.

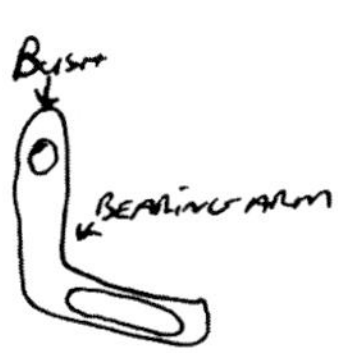

Bearing arm slot for adjustment

The SU Carburettor pipe was cut to give the best fit, as close to 90° as possible and welded. The '80' piece of pipe was cut and swaged to fit into the SU Carburettor pipe. I used blind nuts to fix in place.

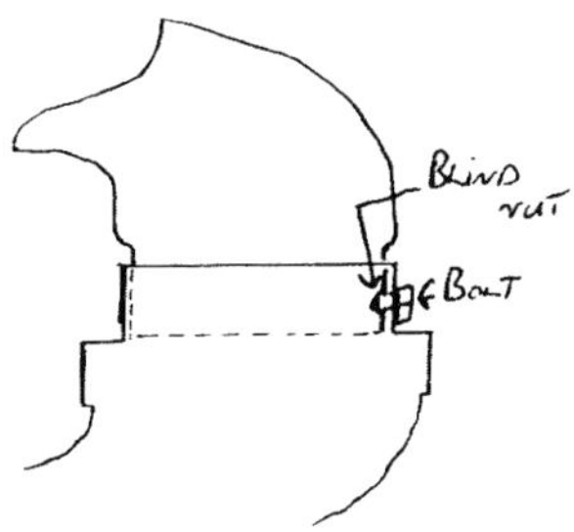

The flexible hose is too short to reach the carb so a piece of 3" tube was used to lengthen this.

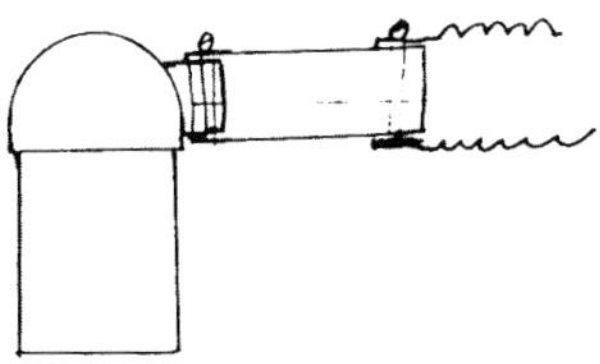

I cut the tube and closed down one end to fit inside the flexible hose and closed the tube down to fit outside the Air Filter. Once I had determined the sizes the tube was drilled and riveted. Then sealer was applied.

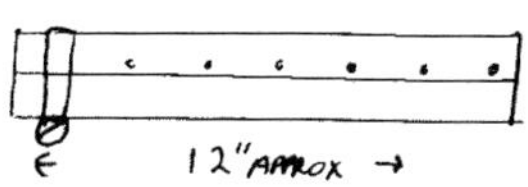

Leave out rivet for jubilee clip to clamp.

Pat Moores ***November 1990***

THE WAVERING SPEEDOMETER – THE TRUTH!

I was following the sorry tales and experiences of people with this problem and was actually dismayed having read an article from some expert advocating change of the speedo cable. The greasing or changing the cable will alleviate the wavering for a period of time but very seldom cure it. The reason is that the cable itself is not the cause of the problem, it is actually suffering from it! For the speedometer needle to register properly the two ends of the cable must run at the same speed (rpm) at any time while the car is rolling, any increase or decrease at the bottom (driving) end must be picked up at the top end, which in turn rotates the speedometer.

If and when a load/braking effect is applied to the top end, the cable will twist, and run slightly slower than the bottom end until it can twist no more, then it will spring forward, slow down again, spring forward etc. hence the wavering!

The cause of the extra load/braking effect is inside the speedometer unit but not the speedometer itself, the culprits are the two mileometers. Each one is driven from the main shaft by a worm-geared eccentric stepping lever. The drying out of lubricant in these two small components create the actual braking effect and the remedy is quite simple:

1. Remove speedometer taking care of the rubber ring behind the chrome bezel rim.

2. The chrome bezel is 'bayonet' fitted, looking from the back one can see which way to twist it for removal. When it is free the double glazed dial and a black surround will drop out with it.

3. Remove the two screws from the back of the unit and carefully withdraw the instrument.

From this point onward take care not to damage/bend the needle or the black face plate.

4. Remove the rubber seal (if fitted) from the back and the two gear wheels will be in view. At the outer end of each of these there is a curved, escutcheon fit spring leaf, holding the push lever and the shaft in place.

5. Press down the little 'mushroom' until the spring is flat and run in a small amount of 3-in-1 oil behind the gear wheel. Release and press again to work the oil in, add some more if necessary until movement is free, repeat on the other side.

5B. If there is no improvement the shaft will have to be removed for cleaning. CAUTION; Adjacent to the gears is the outer rim (aluminium) of the speedo movement, in no way must it be pressed upon!

With the aid of a pair of tweezers or small sharp nose pliers lift up the end of the spring leaf where the small hole is and gently push it towards the centre to disengage the escutcheon.

5C. Remove the lever gingerly so that the spring adjacent to the ratchet will stay anchored and just leave it hanging.

5D. Remove the small shaft with the gear wheel, clean the bore and the shaft with lighter fuel, methylated spirit (NOT WD40) oil it and reassemble.

6. Run small amount of oil behind worm drive and also the outer end of the square drive.

7. Replace rubber seal at the back, insert unit in the casing, fit two screws.

8. Fit black surround and double glazed dial with the small cut-out to the peg on casing.

9. Refit bezel, replace rubber ring behind bezel, refit Speedometer Unit.

To give you some idea of the duration, the first attempt should be less than an hour and a half, with some luck 30 minutes, most of which will be taken up by the removal and refitting of the speedometer to the dashboard.

Tom Tottis ***July 1991***

FITTING FLASHING INDICATORS TO PRE 1955 REAR LAMPS

In reply to Mark Kelly's request in last issue I herewith describe how to establish flashing indicators without making any external differences to appearance of the car. Please note that I only have flashing indicators at the rear. Because I could find no suitable lamp or bulb for the front that didn't spoil the 'originality' – that is white reflectors and the torpedo sidelamps – I perhaps should replace the white reflectors with late type (i.e. 1957 on) sidelamps with clear glass in the reflector position as Carl Haworth has done on his excellently restored 1953 '75' seen at Burghley House. Whichever you choose, the trafficators can still be used concurrently. The problem with only having rear flashers working is that the flasher unit needs to have two 21 watt bulbs drawing power from it to work properly. They will only flash dimly and very fast if one bulb is connected. So you need a redundant bulb for each side *and* a flasher unit for each side. Hiding these is not too difficult in the Rover boot compartment. The early type rear lamp has provision for a second bulb to encompass the lighting regulations for the American 'Dollar Area' export vehicles. On home vehicles the hole only is there and needs a suitable bulb holder fitting into it. I used holders from old broken rear lamps that I had in my collection of bits prising the bulb holders out, fitting them into the holes in the good lamps, pressing back the flanges and for good measure putting a dab of solder on the joint to ensure good contact. Whether the holders are two contact or one doesn't matter as you simply have to fit the correct bulb and run the wire to the appropriate contact – so stop-tail bulbs could be used as well as just single filaments. Incidentally – after making successful additions to the interiors of your lamps it is a good idea to paint the inside of the lamp body white and the bulb 'bracket' as well to improve brightness. They were white when new but are usually rather faded. So on to wiring. Using the single connection on the trafficator unit as your feed you can thread the wire over the door to appear at the rear of the rear door. Remove the lower quarter trim panel and a handy grommet will be revealed which is your route through to the boot compartment each side. To run wires forward should be a similar exercise – the wires appear inside and to the right of the fascia panel for the R/H side and behind the interior of the glove box for the L/H side hence through the bulkhead to frontlamps. In the boot compartment a good spot to hide a double bulb holder and two flasher units is underneath the filler pipe area. Make sure all connections are tight and wrap up with insulating tape for safety – and there you are – the only drawback is having these redundant bulbs lighting up the hidden recesses of your boot compartment, but it's better than having another car's sidelamps doing this for you.

*The advantage of having indicators in modern traffic is an obvious one as motorists these days can't be expected to see trafficators I suppose. It is good however to have them in operation just to show these modern cars how to turn a corner properly!

Matt White ***July 1991***

FLASHING INDICATORS ON EARLY P4s

The vexed question of how to discreetly include flashing indicators on early P4's has been posed many times, and our editor's modifications to his own 1954 '90' in this respect are well known. Without wishing to upset him, or anyone else who has done work in this area, I would like to submit the following solution to the problem, the result of which would be that one could, at the flick of a switch, change from trafficators to flashing indicators and vice versa. As relays and modifications to the wiring are involved, however, a certain fundamental knowledge of car electrics is needed; otherwise a local auto electrician should be able to do the job without much trouble. The modifications could be removed at a later date, if needed, leaving hardly a trace.

There are several problems to overcome:

1. Switching arrangements.
2. Incorporating indicator lamps.
3. Incorporating a warning lamp.

Switching arrangements

The problem here is that three circuits need to be switched over together. Firstly, the feed to the trafficator switch needs to be diverted to go through a flasher unit, as on later models. Next, the current from the trafficator switch needs to be diverted to the indicator lamps instead of the trafficators (one switch for each side). One could just use three switches, of course, but a much neater arrangement would be a bank of three relays, mounted out of sight behind the dash panel, and operated by a single switch.

For those not familiar with relays, and I apologise for boring those who are, they are basically a sophisticated switch, capable of carrying quite a high current, whilst only needing a small current to operate them, a bit like a solenoid. The relays involved here, however, need to be of the "changeover" type (my terminology), as opposed to the straightforward "on-off" type. They are used in many applications in modern cars, so shouldn't be hard to track down.

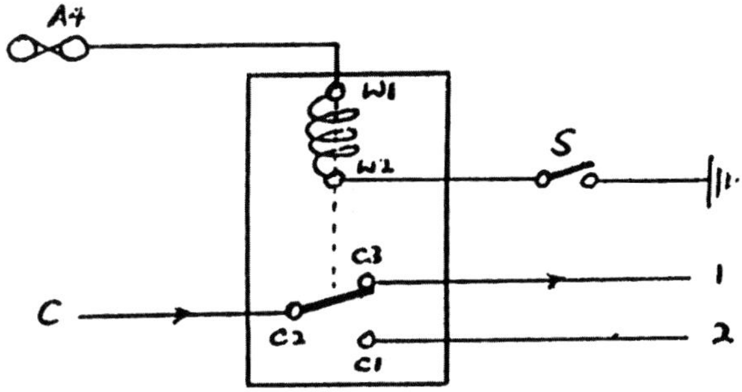

A

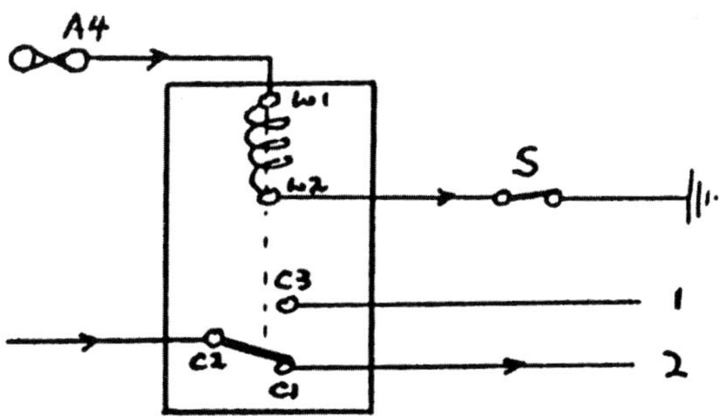

In illustration “A”, with the relay off, the current “C” flows through it via terminals CS/C3 to circuit “1”. When the relay is energised, however, (illustration “B”), by operating switch “S”, the current is diverted via terminals C2/C1 to circuit “2”. Thus it can be seen that not only can a simple “on/off” switch be used to perform the task, but that same switch can operate several relays simultaneously. The switch could, for example, be a normal P4 foglamp switch, mounted on the small auxiliary panel that sits next to the choke control; this would not look out of place and with the relays and extra wiring hidden behind the dash-panel, no-one would be any the wiser. This may not apply to 1950 “square instrument” models, with which I am not familiar. I should mention that the illustration above is for negative earth car; for a P4, which is positive earth the connections to terminals W1 and W2 would need to be reversed.

Incorporating indicator lamps

This is, perhaps, an easier problem to overcome, as cars of this period exported to U.S.A. were already fitted with flashing indicators, so all that is needed is to change the front and rear lamps to U.S. specification.

The front sidelamps need to be removed and replaced with twin-filament bulb holders, together with the more protruding lenses which allow for the larger 6/21 watt bulbs (1954 cars only need single filament bulb holders). The lamps, when installed, are very discreet; how many, for example, have noticed that OLT 399, featured on the front cover of "Rover P4, 1949-59" (Brooklands Books), published some years ago now, has these fitted?

As for the rear lamps, some cannabilism is called for! The bulb holder carrier incorporated in the assembly has space for two bulb holders, but on U.K. cars, only one is fitted. What is needed is to take a single filament lamp unit and remove the bulb holder from the flange plate by bending back the tags that secure it; this can then be refitted to the carrier by reversing the process, and finally soldering into position. Matt White has actually done this conversion on his own car, so he's probably the best person to advise.

Incorporating a warning lamp

This is a legal requirement. Whether or not you bother with it, is as they say, "In the hands of the operator"! Assuming you do wish to, there are not too many options.

The ignition warning lamp is impossible to use, due to the circuitry; similarly, the choke and oil pressure warning lamps are unsuitable as they have other switches to go through before they reach earth. The only one left is the main beam warning lamp, which would be fine until the main beam headlamps are needed; you would then have current feeding back into the flasher unit; also, the warning light circuit would try and flash the main beam headlamps in sympathy with the indicators, which I don't think would go down too well! What is needed, therefore, is a separate warning lamp, which leaves three alternatives; (a) fit a separate warning lamp on the dash panel, heater control panel or an auxiliary switch panel – functional, but wouldn't look right, (b) fit a U.S. specification speedo head, if available, which incorporates a warning lamp (pause for laughter), or (c) the best way, I think, and that is to replace the dash panel with a post-54 type, which already has a warning lamp fitted. You could always keep the original dash, should you want to reverse the process.

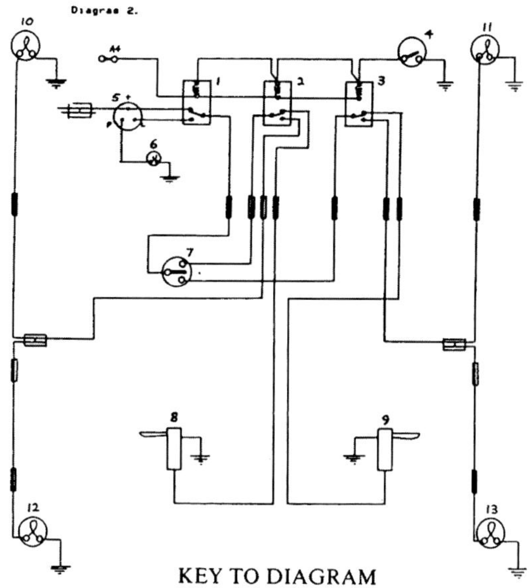

KEY TO DIAGRAM

1. Relay, Feed to traff. switch.
2. Relay, t/switch to ind., L.H.
3. Relay, t/switch to ind., R.H.
4. Switch
5. Flasher unit.
6. Warning lamp.
7. Trafficator switch.
8. Trafficator, L.H.
9. Trafficator, R.H.
10. Indicator, front L.H.
11. Indicator, front R.H.
12. Indicator, rear L.H.
13. Indicator, rear R.H.

High Intensity Sidelamps

I hit upon another idea while I was pondering over the indicator problem. Not only am I not the first one to think of it, however, but it doesn't conform to M.O.T. requirements, as Stan Johnstone found to his cost! I think it's a jolly good idea, anyway, and would be a definite advantage to those who use their P4's on a regular or daily basis. The modification applies to post-56 models and basically involves replacing the sidelamps with twin-filament units, as before. I thought of calling it High Intensity Sidelamps – HIS for short; I suppose you could call it High Energy Replacement Sidelamps, which would be HERS – but I digress. With the lamps on, the result would be very bright sidelamps OR normal sidelamps plus headlamps when driving, and normal sidelamps ONLY, when parked. The lamps would be much more visible than normal sidelamps (about the same as Volvo/Saab daylight running lights) and would really come into their own in poor light situations, like dusk, for example, but would draw less than half the current of headlamps. I am not suggesting that they are a substitute for headlamps however, when these are clearly needed. If anyone is interested in this conversion, I'd be pleased to go into further details; once again, relay and wiring modifications are involved and although, I think, it's a sensible and safety-conscious modification, it's not legal, and would need to be disconnected for the test.

Bryan Plummer ***July 1991***

RESEALING CALIPERS

I have received a number of calls recently from members having difficulty locating the retaining ring in the rubber boot when resealing calipers. When using a screwdriver to install the ring as recommended in the leaflet accompanying the seals, the ring tends to pop out one side whilst being pressed in the other causing much frustration.

I overcame this by making a disc from 16 gauge alloy of a diameter that would fit inside the retaining ring and the ring would not readily drop off. A short bolt was attached to the centre of the disc as a means of removing same when the retaining ring was finally in place.

Proceed as follows:-

1. Fit piston sealing ring in groove in caliper.
2. Lay caliper on its side.
3. Place ring on disc.
4. Insert disc and ring upwards into boot.
5. Lay boot over hole in caliper holding disc and ring clear.
6. Work boot into position with remaining fingers, followed by disc and ring – feeling through the upper side of boot.
7. When disc is centred, press home then remove disc leaving ring behind.
8. Check for correct fit.
9. Lubricate piston with Girling rubber grease or brake fluid and insert through boot.

Early pistons use the smaller ring to retain the boot in the piston groove – later pistons do not.

By the time you have fitted three boots, the fourth should go in at the first attempt.

Derek Humphreys ***January 1992***

ROVER '80'/WEBER SINGLE CHOKE CARB CONVERSION

The Weber takes a 'push on' petrol-pipe, so I have fitted 5ft of 'Gate 3225 Multi-fuel' ¼in universal fuel line and fitted in-line filter supplied with the Weber carb in place of the Rover blackplastic fuel line. (The filter should be fitted horizontally)

Now to Commence!

Remove the ball joint and bracket fitted to the new Weber butterfly shaft and RETAIN the ball and nut.

Remove the old Solex carb flexible shaft (item 46 on page 163 the Parts Book). Fit this to the Weber (it will fit O.K.).

Please Note

Due to the fact the new Weber carb butterfly shaft opens in the REVERSE direction to the old Solex you must now proceed as follows:-

Disconnect the 2 orange and 2 slate wires from the cam switches. Detach the cam switch assembly from the bulkhead by removing the three ⅝ x ¼ BSF bolts (don't drop them!).

Release the switch locknuts and unwind them to the ends of their threads. Give them a good clean with electrical switch cleaner and make sure they are working properly.

Remove the cam from its shaft (note which way it faces now, with the toe of the cam uppermost, as per fig C32 on page C.28 of Workshop Manual). Check condition of shaft and fit a new bearing (Fig 35, Page 163 Parts Book) if found to be worn.

Make up the conversion plate as per diagram and fit the new Weber ball joint as indicated in diagram No. 1.

The old Rover ball joint is no longer required, but do not cut it off as it acts as a retainer for the conversion plate (see diagram 1).

Drill the pivot shaft bracket and fit a new ½in x 3⁄16in bolt, SPRING-washer and bolt as indicated on the conversion plate. NOTE: This plate MUST be secure.

Rotate the pivot shaft (item 23, Parts Book) 180° from its original ‘at rest’ position so that the ball joints now face the engine valve cover.

REVERSE the cam on its shaft rotate it 90° so that the slot now faces UPPERMOST with the locking bolt horizontal. (You may need to reverse the direction of the 1¼ x 2BA bolt to assist tightening finally).

Tighten the cam as indicated and adjust the switch clearance as per Workshop Manual, but remember, THEY ARE NOW WORKING IN REVERSE.

Fit the cam/switch bracket back on the bulkhead and correctly align the Weber carb shaft and pivot shaft assembly for COMPLETE FREEDOM OF MOVEMENT. Tighten all bolts gently.

Reverse the throttle lever (Fig 28 Parts Book) on its cross shaft, so the ball now faces the cylinder block.

SOLEX/WEBER CONVERSION

Diagram 1.

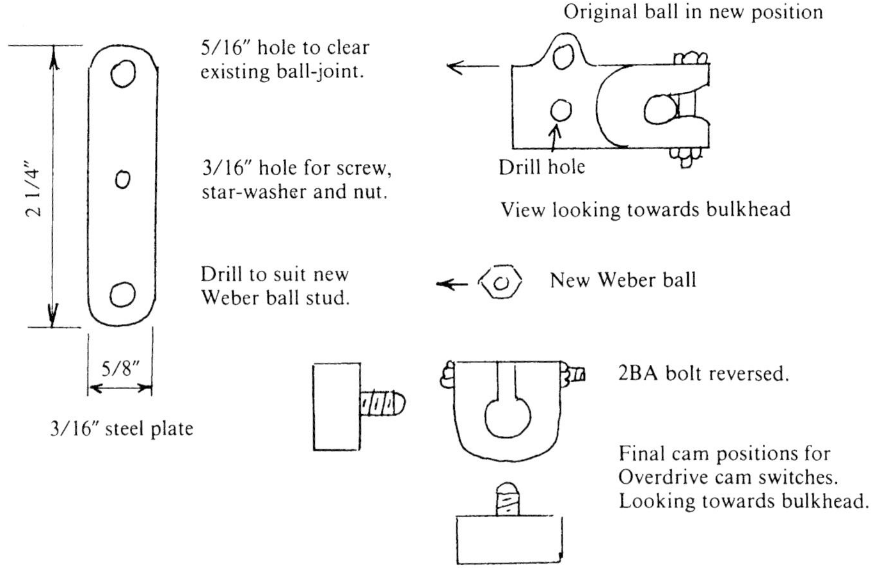

Set the collapsible shaft to 9³⁄₁₆ between ball centres (as per paragraph 13 on page M.1.27 of the Workshop Manual).

Fit the collapsible shaft to the new Weber ball joint and to the bulkhead cross shaft leader assembly, tighten the lever assembly.

Check the cam switches are operating freely and fit the Slate/Purple and Slate/Blue wires to the HORIZONTAL switch and Orange/White and Orange/Black to the VERTICAL switch.

Remember the switches are now reverse in operation.

Check electrical circuit now using a 12 volt test lamp.

Re-direct the choke cable under the heater blower motor and fit to the carb. (it should just be long enough!).

If too short, make up a metal sleeve and fit this over the outer cable and clamp this to the Weber bracket.

You may find that the flexible shaft coils tend to open up as it is now working in the opposite direction, but I have had no serious problems with it.

I fitted this carb conversion in July 1987 and am very pleased with it.

Now the important bits!

Performance – A much smoother response than my old Solex carb (both the modified and unmodified types) no 'flat spots' whatsoever.

Fuel Consumption – About 25-30 solo (20 with Solex) and about 20-25 when towing my caravan (15 mpg with Solex).

My best 'Top up tank' to 'Top up tank test' was nearly 34 mpg (driving very carefully I might add!).

A price quotation on this conversion package can be obtained from Mr Richard Findlay, Land Spares, 3 Woodside Parade, Woodside, Leigh-on-Sea, Essex SS9 4SS. Tel: 01702 512110.

Parts Needed

Weber Carb Type 34 ICH (2/350)
Carb to Manifold adaptor assembly
¼ BSP Blanking plug for adaptor (if required).
(Some adaptors come with a hole drilled in them)
Insulator pad (Carb to Manifold) if required
2 Gaskets for above
Push-on type sleeve for vacuum pipe
* Air Cleaner to Carb Elbow (steel)

* Rubber Hose connector
* 2 new Jubilee clips
* Air cleaner flexible hose (pattern type)
* these 4 items are not essential, but the old Solex carb. took a larger diameter hose. This can be tightened up more, but does not fit so securely.

Malcolm Belchambers *March 1992*

HINGE AND BRACKET

Cease reading at this point if your car is in concours condition. This article is aimed at owners of cars with doors that appear to weigh 4 tons each, these doors being opened only by sinking one's heels into the ground and yanking at the door handle with a bionic arm. At the very least the handle will remain attached to the door, otherwise prepare yourself for a hard landing (very amusing if you're watching).

Seriously though, folks, spare a thought for your poor old door hinges which have possibly been faithfully opening and closing regularly for up to 30 years.

A few minutes maintenance on a regular basis will ensure they go on working for a further lengthy spell.

On pre P4 Rovers, as with many pre-war and some post war cars, the door hinges were situated on the outside of the car and were thus easily oiled. Neglect caused not too many problems as the bearings and pins were easily replaced.

Changing fashions saw the disappearance of running boards, separate headlamps and outside door hinges. The new flat sided cars had an advanced (?) design of hinge. Advanced design always brings inaccessibility – or so it seems, the hinge now being hidden from sight. Therefore let it not be a case of out of sight out of mind on P4 hinges.

A light oil regularly applied to the groove in the top of the hinge arm should make its way to the pivot. A seized hinge, which occurs mainly on rear doors, can usually be freed using an oil spray into the hinge opening, and the door worked backwards and forwards (that's where the bionic arm comes in useful). Several doses of this treatment may be required for a bad case. Ignoring the hinge over a period will eventually result in hinge failure and consequent dropping of the door. Changing P4 door hinges isn't a pleasant occupation unless you like grovelling around the floor, so out with those oil cans now.

Now on to the bracket, for want of a better word and because it fits in with the title.

Do your doors protest with an agonising creak when they are fully opened over the last two inches or so of their travel? Well, the remedy for this common ailment takes slightly longer than hinge oiling, but it is a job worth doing.

To cure this appalling noise, which wouldn't be out of place in a Hammer movie, first remove the inner door panel, or at least far enough to gain access to the hole adjacent to the door check strap. This hole was originally covered by a square of black waterproof material. If it is still there, there's a good chance the panel hasn't been off before. Anyway, remove the waterproof square, underneath through the hole will be seen the door check strap and two spring brackets riveted to the door, one each side of the strap.

These brackets will usually be found to be very rusty, their role in life being to retain the door in the fully open position, being forced apart by a spacer on the check strap.

A dab of grease on the spring brackets one above and one below the strap will cure your creak. Difficult to write about but easy to do.

Once these jobs are done you won't be embarrassed when opening your rear doors even for the Queen (well Princess Margaret had a P4).

Finally, don't forget to oil the hinges on the fifth door (didn't know you had one eh?). Well it's located under the rear bumper and covers the, often forgotten until it's needed, spare wheel. Incidentally, ignore the last item if your car hasn't got a spare wheel door. Some have rusted away or fallen off, but I must assume that the other 4 doors are still there, if only to keep the draughts out.

Colin Blowers ***1985 Yearbook***

REAR HUB OIL-SEALS

It seems to be a widely held view that special equipment is required to remove rear hub oil-seals on P4's.

The autobook manual states that special tools are needed to service both bearing and oil-seal. This is not the case. The Rover manual is also not clear on this point.

The Rover tool No. 275870 is only required to separate the whole assembly. If everything else is in good order, the seal only may be changed quite simply.

Reference to Fig. E-3 (Rover Manual) shows that the seal is pressed into a sleeve in the housing and contacts the bearing securing sleeve. With the assembly removed from the axle it is simply a matter of pulling out the seal. "How?" you cry! The following tip was given to me many years ago by an old Rover storeman.

Drill 2 or 4 equally spaced holes in the outside metal part of the seal. Insert self-tapping screws (NOT countersunk head) into the holes. Take great care to avoid damage. Lever under, or pull on, the screw heads and the seal may be fairly readily extracted.

The new seal can be gently tapped into place with a hammer and drift.

The seal is an ordinary 2¾" x 1⅞" x ½" seal and is easily obtained from a bearing/seal specialist.

I have carried out this job on a number of cars with complete success. Failure of these seals seems to be a common problem, so I don't think that fitting a second-hand unit would be worth all the extra work involved, only to risk another seal failure in the near future.

The rear hub oil seals never seem to last for a particularly long time but they are not difficult to replace. Harry Stevens is under the common misapprehension that a 20 ton press is needed to remove the seals, but in fact this is only needed to remove the collar retaining the bearing. The oil seal is in its own collar which can be removed, either by strong fingers, or carefully using a pair of pliers to overcome the oil suction, which can be quite strong. A new seal can then be put in and the collar pushed home. The seals only cost about 70 pence and it takes me only half an hour to do the whole job for one hub. It would therefore seem extravagant to obtain a complete replacement drive shaft from a breakers' yard.

David Messenger ***1985 Yearbook***

CORROSION AND THE P4

The conscientious owner wishing to preserve an older vehicle, will be only too well aware that the exposure of steel cars to a damp environment inevitably leads to corrosion. However well the cars are protected when new, eventually there is a breakdown in the paint or underseal layers so allowing water to attack and rust the bare metal; the rust swells, lifting the surrounding paint thereby providing a fresh environment for water to seep into by capillary action; and so the cycle continues. Time and effort are then expended by the owner to limit this damage. Despite the great P4 strength of mechanical longevity, even when harshly abused, bodywork decay has sent many cars to an early grave.

The original purchaser probably did not intend to keep the car for 25 years, therefore it is not surprising that the good advice contained within the Vehicle Instruction Manual was not heeded:-

'It is well periodically to wash the underside of the car, to prevent mud pockets and the consequent tendency for rust formation. The use on roads during frosty weather of salt, sometimes in quite strong concentrations, is now being widely practised. Whilst special protection has been provided for the under surface of the body, due to its highly corrosive nature salt deposits should be washed off as soon as possible by thorough under washing of the car.'

This is from a manual for '80' and '100' models – earlier car manuals lack the sentence on winter salt since that practise only became widespread in the sixties. In fairness to those original owners, the precise locations where problems were to develop is something only hindsight has provided. So where do P4's corrode? What do we look for when surveying a prospective purchase? Which areas need a regular de-clugging?

The damage which occurs falls into two distinct categories; the initial surface rusting is followed by rotting through of the structure.

The Coachwork: Surface Rust

Damage to the surface of the paintwork usually results from abrasive damage, for example stone chips, small knocks and dings. This then gives rise to surface rusting which penetrates beneath the paint surface to spread blotches of corrosion over a surprisingly large area. Suitable treatment with a grinder or sander will usually reveal bright shiny metal, repainting restores the finish. The consequences are not usually serious if caught sufficiently early, but do detract cosmetically, the front valance (the panel below the front bumper; usually never washed) and indeed the front face of the car can suffer badly in this way.

The Underbody: Body Rot

This is often hidden in the initial, surface rusting stages since it occurs in parts of the car that the eyes do not normally reach. Paint apparently blistering from beneath and holes through the coachwork indicate the need for welding and other bodywork to rectify. The rust formed has a larger volume than the preceding steel, which explains the blistering, and has little strength which is why monocoque vehicles (i.e. lacking a chassis) can become unsafe and dangerous. Fortunately a P4 is held together by a robust chassis, so even in cases where the appearance suggests consignment to the scrapyard the vehicle may be quite sound for road use.

Age, storage and use

The type of use, the amount of it, and the storage conditions of a vehicle will determine the nature and amount of the deterioration. Given minimal care by their previous owners, a low-mileage farm car living at the end of a muddy track will typically exhibit more corrosion, and require welding, than a high mileage car with stone chip damage used only in summer and for touring holidays. All P4's are now well past their twenty-first birthday, consequently past and indeed present vehicle storage will affect condition. A vehicle stored in a damp garage will suffer surface corrosion throughout – e.g. even the instrument panel and all interior chrome may fester, rust may also stain the headlining and upholstery. Most of this will polish up well – but what's it like where you can't see? Water vapour carried in the air we breathe is responsible for this passive deterioration – the older the car the greater the opportunity. Thus vehicle mileage is no guide to vehicle rusting and vice versa.

Surface deterioration of the coachwork

Exposure to rainfall, particularly in cars which have been stored outside for part of their life, can lead to much unsightly deterioration which may lead to further problems. so rectify early and prevent future problems.

a) Water collecting to the rear of the front apron (the horizontal panel between the bumper and wings; '52 onwards) may cause rusting by the joint where the panel enters the front wings.

b) Road dirt and water held beneath the stainless steel side strips dividing the bodies of post-'58 cars permits corrosion, initiated from the mounting clips. Similar rusting due to water retention in earlier cars may be apparent around the rubber base of the torpedo sidelamps ('54 -'56).

c) The aluminium roof guttering would appear to be crimped to the steel roof panel. Flexing of the vehicle causes the paint seal to break thereby permitting electrolytic corrosion between the two metals. The paintwork blisters and lifts; particularly on the descending portion by the rear doors.
d) Water running down the roof over the front windscreen can penetrate behind the seal and cause surface rusting of the bodywork by the bottom corners. Weathered windscreen seals appear deeply crazed.

Water from the front windscreen and front guttering (post '56 cars have an extra lip) is directed beneath the bonnet; drainage channels transfer it to the sides of the bulkhead. The water accumulates on a shelf, attached to the side of the bulkhead. There are two indents on this shelf permitting water to drain down the scuttle sides and out beneath the car through official drain holes. This pathway easily becomes clogged with silt. When travelling forwards water may slop over the rear edge of the shelf and down the channel between the plate locating the door sealing rubber and the angled bracket to which the front wings are bolted (i.e. over the vertical row of bolts). The lower door hinge is protected (by a piece of leather-cloth – '55; or a metal shroud post '57) and the water drains out of a small gap which can be discerned with the car door open; i.e. between the door rubbers and rear of the front wing. However, if the car is stationary or sloping downhill the water may run down the inner wings on top of the chassis in the direction of the steering idler mud trap (see later). Alternatively it may run out beneath the rubber sponge sealing the gap between the top of the car body wings and the scuttle. Experiment for yourself with a glass of water, if no water emerges beneath the car then it's time to start worrying.

If any of these elegant pathways get clogged then water goes where it shouldn't; down inside the scuttle, into the door hinge area, under the rubber sill mats. Indeed water may be found standing on these sill mats after heavy rain (blown in past the door edges) and it may then creep under the mats to soak into the beautiful absorbent door seals etc.
e) The large three piece rear windscreen (post '55) does well to stay leakproof. The sealant used to join the glass to the rubber may crack and permit leaks at the bottom corners where water accumulates (just by the base of the two vertical strips).

Once past the seal the water runs down behind the boot trim, it may appear initially around the shock absorber inspection plates. Larger quantities can end up at the tail end of the boot corroding the floor and working up into all the beautiful sound deadening felt. This process is concealed by the boot mat.

f) Water draining down from the rear of the roof may find its way into the boot drainage channel. The base of this may be uneven so helping the collection of leaves, pine needles and other rubbish. The paint may peel, rust may develop round the edge, especially the top corners (where there are welds), and in extreme cases small holes may permit water entry into the boot. Water should drain down the channel and away, unless it is waylaid by the lower boot sealing rubber. Like so many of the seals on the P4 this is composed of sponge rubber, and when damaged absorbs water permitting further corrosion (and occasionally rot) beneath the seal.

g) The doors. Water draining inside the doors should be able to get out. If it can't, then the interior door mechanisms (locks and winders) deteriorate and damp can then recondense throughout the interior. On steel doors (post '63) rust may break through on the outside above the bottom stainless moulding and also beneath the door.

The drain holes at the base of the doors should allow water to pass into the stainless moulding. These have small cutouts at the extreme ends, on their inner surface, for the water to drip away. Small amounts of dirt easily foul up the system. The door drainholes are not exactly in the door base, this permits water to creep between the doorskin and frame (the process which condemns steel panels since such a region is never well painted). Neither is the area well ventilated. Often the stainless moulding retaining clips (steel) are corroded within the channel.

Aluminium doors do not rot through, but they do corrode – typically water creeps behind the paint, which has difficulty in adhering to aluminium. Uneven, broken and blistering paintwork then develops above the stainless trim and on the underside of the door behind it. Corrosion may also occur on the forward edge of the rear door swage. The electrolytic action of winter salt is particularly harmful in promoting corrosion. However, unlike steel, the damaged paint should clean up satisfactorily.

h) Similar problems are apparent in the fuel filler door (post '54) where the drain hole is situated above the rusting line. The rear number plate plinth can also suffer when water gets trapped between itself and the boot – later cars (post '56) suffer with the p.v.c. sealing strip which limits ventilation.

The rear bottom edge of steel boot panels suffers likewise from condensation.

Deterioration of the underbody

Abrasion by stones will damage the original protection; the resultant surface rusting sweats away concealed by the accumulation of moisture retaining road silt. Winter salt can also be harboured in these death dealing poultices. The corrosion eventually eats through the underbody to the coachwork.

The original underbody protection was only adequate; early cars seem to be finished with primer and chassis black. Later cars may have this supplemented by factory applied underseal. The actual chassis of the vehicles seems only to have been covered in one coat of chassis black. Since these protective layers suffer much abrasive damage, deteriorate with age, and were seldom, if ever, cleaned, it is not surprising that most outwardly shiny P4's have rusty underbodies.

Beneath the front wings

The front valance panel; this runs between the front wings, below the bumper. The construction of the joints permits water retention by capillary action and in extreme cases may appear holed.

Under the wing, mud accumulates on the ledge below the headlamp bowls, which the valance and front apron are bolted to. In extreme cases, the base of the front wings rot through, especially by the bottom corners of the radiator grille. Damage may also extend to the inner wing and base of the radiator support frame. This area is fed with water from the bonnet drainage channel drain holes.

Mud collecting around the headlamp bowls will eventually rot them through. This leads to a deterioration in headlamp reflector silvering and perforation allows condensation to form within the headlamps..... (brings tears to my eyes!). Water may also penetrate behind the seal between the headlamp and the front wing leading to lacey edges around the headlamps mounting area.

Mud may accumulate above the headlamps leading to holes through the wing on pre-'57 cars.

Post '57 vehicles, with the indicator at the top of the wing, suffer silt accumulation which initially rots through above the lamp and thereafter all round it.

Mud may also collect around the front wheel arch lip leading to frayed edges and holes through the swage line. However the main weakness, without exception, is the rear of the front wheel arch where the wing is sealed to the bulkhead. Corrosion travels along the base of the bulkhead in the general direction of the front door, weakening the whole area behind it. Why?

The gap between the bulkhead and the front wing is sealed by a rubber strip attached to a metal J-shaped securing plate and fixed to the bulkhead with set screws. The rubber strip is compressed against the bulkhead to form a seal and at right angles pressed against the inside surface of the front wing. Thus there are several layers of material for water to seep behind, and once the metal J-strip corrodes through (from the bottom upwards), filth can more easily penetrate the void behind. Thereafter the rot eats the base of the bulkhead and the wing behind the stainless moulding. A line of corrosion may be apparent outside above the moulding and up the wing in line with the rubber seal. Corrosion may progress into the door hinge mounting and also beneath the sill rubbers..... eventually weakness may develop in the base of the B/C post. The wheel arch lip is usually less badly affected except at the bottom corner, by the rubber strip, which stays damp longer because that's where the water drains.

Immediately above the J-shaped rubber strip is a piece of sponge designed to stop large amounts of road dirt entering the engine bay. Eventually this rubber saturates and prolonged contact between steel and damp rubber leads to corrosion, water then appears to be fed into and along the spot welded join between the front and inner wing panels causing severe rust induced swelling in extreme cases.

The Chassis

Mud can accumulate on the top surface of the chassis beneath the front inner wings, particularly between the bolts securing the steering idler (see Overdrive 43, 13), the chassis may even be perforated. (The steering box side seems to benefit from slight oil seepage) Dirt and small stones may also collect at the bottom of the road springs and annoy the steel base plate.

The jacking point supports are weakened by exposure to the high velocity fire of road filth spraying at them from the front wheels – as is the handbrake mechanism protection box. Corrosion may be sufficient to cause these items to break away and rot may nibble the floor around the handbrake detectable inside the vehicle. Contributions to this latter also come from a failed bulkhead area.

The two main chassis members running under the car are usually acceptable; due in part to the covering of oil received from a leaky rear main oil seal via the bell housing base. However the chassis outriggers, (extensions supporting the jacking points and body mountings bolted to the sills) are more exposed and may be eaten away if filth has built up in them.

At the rear of the car, the sides of the chassis are exposed to abrasion from road dirt in the rear wheel arch, the petrol pipe on the top surface may become fragile and the rear axle bumpstop rubbers may drop off when rust penetrates their mounting. The rear body mounting support channel may become weakened to the extent that the rear body is not located to the chassis!

Lastly, spray may consume the spare wheel tray and the inside of the spare wheel door with surface rust; however these items may clean up satisfactorily. The battery box is more likely to be nibbled by escaping acid than rust.

Rear Wheel arches

Construction of the wheel arch lip is more complex than at the front; the inner wheel arch is spot welded all the way round to the outer wing. Water penetrating the join will cause the lip to swell; accumulated mud on the lip will preferentially attack the spot welds (weld material is always more susceptible to rust attack) and rust may break through the coachwork in a line around the arch, especially behind the D post. Here mud collects, travelling up the wheel arch from the bolt securing the stainless moulding; the inner arch may become perforated so allowing water into the rear door hinge area and beneath those sill mats once more; the lip by the moulding may disintegrate. If the inner arch gets perforated further forward, then corrosion will break through the coachwork down the D post – but this is extreme. The inner surface of the inner wheel arch may rot around the boot hinge mountings (boot refuses to stay up). Other holes may permit water to enter the car interior (found beneath the rear seat cushion) and the boot etc.

Splash plates fitted in the rear valance do a good job of preventing muck from rotting out the spare wheel door hinge supports; but they are usually sacrificed in the process and eventually holes appear by them in the valences. Absence of the splash plate also provides less support for the spare wheel door hinges which are then prone to seize – thereafter corrosion and excessive leverage on the door will cause the hinges to break away. In certain cases it is not unusual to find little correct coachwork on a P4 beneath the rear bumper line. The boot floor corners (above the splash plates) are also prone to disappear – particularly on the petrol pump side (not '80' models), water is then able to enter the boot, and escape this time, but the damage has already been done.

Summing Up

This may all sound depressing but merely serves to illustrate the downhill path; fortunately little of the above will put a car off the road. The areas of interest to an M.O.T. inspector are the security and condition of the lamps and likewise for the body mounting points; otherwise structural integrity is maintained by the chassis. However, perhaps it will be your own inability to continue with a suffering, sad looking vehicle in the long term which sorely taxes your staying power, even when road-worthiness is not the issue.

This list is only a guide based on my experience; corrections to both errors and ignorance would be appreciated.

Paul Hart ***1987 Yearbook***

CLUTCH SPINDLE OVERHAUL

With regards to the recent "Overdrive" bulletin, I thought that you might like to know about my recent experiences with Rover spares.

Firstly, the Windsor Motor Company of Sheet Street, Windsor, have been having a stores clear-out and recently had on display a brand-new distributor for a Rover '75', and a brand-new windscreen wiper motor for a Rover '90'. (This I believe, fits quite a lot of the P4 range). They also had some rear number-plate brackets, also brand-new, at £1.00 each.

Over the past several months the clutch on my Rover '100' has been becoming increasingly stiff in operation, being due, I thought, to wear in the bronze-type bush which connects the clutch pedal spindle unit to the gear-box casing. These bushes are very easy to replace once you have got your car over a pit or on a hoist, and are still available, I am informed by a Rover dealer's storesman, at 15 pence each plus V.A.T. This is what I paid. The current part number is: 217984. They are also available from a dealer in 'Lewisham' who has had some bushes specially made, and they come with two small sandwich plates that go on either side. They sell these at £4.00 each plus 20 pence postage.

Now back to my clutch..... I found that the actual clutch spindle, which is located through the main chassis member, was almost completely seized. It can be seen from the Workshop Manual that the brake pedal is located on the off-side end of this and pivots freely on the end, retained by a small bolt. The whole clutch linkage and brake pedal should be, and was, removed allowing access and a judicious tapping with a hammer to knock the clutch spindle out. I found it had not seen much lubricant for some time, and was partially rusty. A good clean with emery paper and plenty of grease meant that on re-assembly, the clutch works beautifully.

This is quite a straight-forward job and certainly makes a great difference to the feel of the car. As with any car maintenance, a good look at the Workshop Manual is essential before starting this job, and then it all makes sense! Incidentally, I have not seen any particular reference to routine lubrication of the offending clutch spindle, and it could quite easily be missed. A good dousing with oil every few months should do the trick.

I hope that my experience may be of help to others

Keep up the good Rover work

Andrew H. Shaw ***1988 Yearbook***

SYNTHETIC ENGINE OILS?

No one seems to have objected to my putting SAAB SAE 75 oil into my gearbox, so how about a synthetic oil for the engine?

Don't get me wrong, modern oils like Castrol GTX and Duckhams are very good – but give them the finger and thumb test after a year and 5,000 miles and there is a noticeable decrease in slipperiness.

I first tried synthetic oil in a VW Golf five years ago: that car has now done over 100,000. My 120,000 mile SAAB Turbo has always had synthetic: it still runs like new. Our Morris 1300 (66,000) and Morris Ital (120,000) run noticeably better with synthetic, so why not my 1955 '90' at only 110,000?

I knew my engine was clean inside, since I had dredged at least ¼" of gunge off the bottom of the sump last year. Previously I had run some highly detergent diesel oil through the engine for a week – that had certainly shifted the muck.

Auntie has some high-friction areas – the camshaft in particular – surely synthetic can't do any harm, except to the bank balance? So in January, 8 litres of Mobil 1 was installed and as a result the engine starts easier and runs noticeably smoother and freer. Oil consumption appears unchanged at 200 to the pint as always but I appear to have 'lost' about 5 psi oil pressure when hot and I can hear that rumbling main bearing slightly more – it will need attention shortly!! Suppose I shouldn't grumble after 37 years?

All very well having an oil gauge, but, rather like a heart monitor, one soon becomes a hypochondriac … but where *did* that 5 psi go? Lurking deep in the engine compartment is the oil pressure relief valve – adjustable on early cars, fixed on roller-cam models – but after 37 years the spring was unable to maintain even 40 psi when hot. I could, of course, screw it all the way on and jam the ball on the seating, but the cold pressure would then be astronomical.

I invested in a new spring: it was longer and stronger than the old one and the 'hot' pressure shot up to 50 psi whilst the 'cold' pressure remained at a reasonable 65 psi. If anyone is interested, I took a set of measurements for comparison purposes …

In April it was colder than January! As I write in May it is roasting – I see no reason to change my conclusions that synthetic oil has been an improvement, and they reckon it can stay in for 25,000 miles so it should work out quite a bit cheaper in the long run. Whether it is worth the initial £40 investment I shall leave up to you.

Jim Butterworth ***September 1992***

ROVER '60' VYR 720 (ALIAS AUNTIE VERA)

Having owned Auntie now for just over two years, I thought I should put pen to paper about a persistent problem I've had with big end knock on acceleration, coming in just prior to gear change rev band. As this was now to be my everyday car (all this came about, I'm happy to say, when the wife's Allegro needed a new engine and we were looking around for a Rover 213, the same as I was running, when she happened to say: "You've always wanted another P4 like we used to have 20 odd years ago". This was also a '60', a nice black 1956 with old type wings and high boot and wrap around window. I still like that shape. (Anyone out there got my old '60' SLK 967? I would like to hear from you) Well I mean fancy the wife waving a red flag to a bull. You get no points for guessing what car she is now running! I retarded the ignition slightly and drove cautiously for some 14 months before I had time to strip and rebuild the engine. Funny thing was even though it was a big end knock the symptoms were opposite, i.e. labour it on hill it would go quiet as a mouse and any speed over 50 it was OK. It also had a good oil pressure at tick-over when hot.

Eventually I did a complete strip down only to find nothing was worn more than 0.0008 thou (which just goes to show how well built these engines are after 34 years). Despair set in as it was expensive to strip out the crankshaft, as new shells and seals were fitted back. I do not believe in using old ones once disturbed. Anyway it had a good decoke, valves ground in, hydraulic tensioner overhaul, new jet and needle in carburettor, plugs, points, leads, condenser, water pump overhaul and new hoses.

The engine appears to be in good condition as it only uses ¾ pint of oil between services. I do this at 4,000 miles and change the filter every time.

So back to the problem of big end knock. The only thing that can cause timing problems must be associated with the distributor. So a letter to Stan Johnstone explaining the problem and could he give me any information on the correct distributor, vacuum unit, springs, etc. Stan kindly wrote back giving me the correct spring part number and Lucas Technical Information Dept address. Many thanks, Stan, for your interest and information. So a letter off to Lucas with all the details from the distributor and vacuum unit. Lucas's Mr Davenport, Customer Technical Services, was very helpful.

The following information may be of help to others.

Type DMBZ4A 404 10D 6113 10.58

Vacuum 419681 4-18-12

4 = is inches of mercury at which advancement begins

18 = is inches of mercury at which advancement ends

12 = is maximum advance in distributor degrees

The cam inside the distributor body can be assembled 180° apart. One side is marked 13, the other 25. The former is to go against the stop.

The springs are LU410605, both the same, and consist of the following: 5½ turns of 0.026" SWG wire outside dia 0.192", overall length including loops 0.705" – all this exciting information and guess what? All was correct. I even marked the flywheel in degrees and checked with a strobe the vacuum unit. You've guessed, it was correct. I was by now getting exasperated – it has got to be the springs. Yes, they were very shiny for 30 odd year old springs, but did they measure up correctly. Perhaps they are new, but of a softer spring steel. So I took the bull by the horns and took one out and replaced it with a homemade one of thicker gauge. To my amazement it cured the problem and it was advancing fully.

This was left like this until I came to do another service of points and condenser. Whilst undertaking this (fate I suppose) I dropped the unit holding the the wires and spring contact into the distributor body. Whilst retrieving the nut, having removed the base plate to facilitate, my old problem loomed up before me. Why had I not noticed this before? The fixed posts supporting the springs were bent inwards slightly. I immediately bent them to 90º, replaced my homemade spring with the original. *Problem cured.* I can only assume a previous owner or garage mechanic had fitted new springs and bent the posts in to get them on (as they are strong) but not enough to make the rotor arm sloppy in the static position. So all you P4 owners, before condemning the engine to the scrap heap or think of buying a new distributor, *CHECK YOUR POSTS!!*

Mike Coleman ***September 1992***

REAR WINDOW REPAIRS

I have recently had the rear window frame of my '100' rechromed and have renewed the rubber seal so I thought members might be interested in this operation.

The Workshop Manual details how to remove the window from the car. However, when all the clips were removed, it needed levering from inside with a tyre lever to unstick it from the body and two helpers outside to catch it.

Clean the old sealer from the frame and remove the old rubber seal. I found the best way to do this was to cut down each side of the seal with a Stanley knife, after which it comes out in three pieces. Remove the countersunk screws from each side of the frame and after a considerable amount of pulling, the frame will come away from the glass. A thin blade can be a help here to unstick the glazing rubber. New glazing rubber will be required and I used a channel section, catalogue No. R315 from Woolies of Market Deeping.

The Works Manual again explains how to fit the glass into the frame which was fairly straightforward. However, I was unable to close up the two sections of the frame to insert the screws in spite of using cord loops tightened with lengths of dowel to pull them together. I therefore retired to the local windscreen fitting depot where it took four of us two hours to persuade the frame to close. The rubber seal is then pressed into the frame with the aid of soap solution and a screwdriver.

Not having the clamps shown in the Workshop Manual to pull the unit to the body, I laid it over the aperture and then connected a set of pulleys to the centre bottom of the frame with the other end anchored to the hole in the transmission cover through which the gear box is filled. This enabled the bottom of the frame to be pulled in and a pair of clips loosely fitted to the bottom edge. The pulleys were then transferred to the top and clip fitting continued around the screen with the additional aid of a tyre lever to pull the frame in close to the body.

Screws can be finally tightened when you are happy with the alignment. Two lengths of self tapping screws are used and care needs to be taken to ensure that the correct length screw is used in each hole so as not to foul the outer skin.

Shims are also fitted under some clips but they do not necessarily go back in the same position when a new rubber is fitted and additional shims may be required. Do not use the screws to pull the frame into place – lever the frame then tighten the screws.

I did not use any sealant between frame and body before fitting but subsequently injected sealer under the rubber.

In hindsight, it may have been the pre-formed rubber channel which caused the difficulties in closing the frame, perhaps the flat rubber glazing strip, 2" x 1/16" – 11' required, which Woolies also sell, would have been better as this is of slightly thinner section.

My final advice is that unless things are extremely desperate, leave well alone.

D. Humphreys ***March 1993***

S.U. FUEL PUMPS

On repairing a S.U. fuel pump recently, I found that I couldn't get it to run properly. I then remembered (terrible memory these days!) that about three years ago, a batch of contacts were made with the "roll-over" springs just too long, making it difficult for the points to trip over without the springs getting stuck under the 'cage'.

I won't mention the company, but the contact set is sold in a small blue and yellow box. I think you know the ones I mean.

If you have removed a set of contacts that have just burnt out, but have been working well, you can very carefully remove the offending springs from them and place them in the new cage, with gentle use of a pair of long nosed pliers, but I wouldn't really recommend this unless you are desperate. Ideally only use S.U. points in the plastic bag. Single pumps were SU AUA 6021 and double ones SU AUB 6022.

Don't forget to bleed the pump of air on fitting by 'cracking' the outlet pipe with the ignition on, allowing the air to escape and then a steady stream of petrol.

If after doing this, the pump is still rattling away, you very likely have a tiny hole in the inlet pipe, which will have to be renewed.

Stan Johnstone ***March 1993***

CHANGED EARTH POLARITY

From the early 1930's onwards British cars were all built with what is known as positive earth, that is, the positive side of the battery is connected to the car chassis to form the return path for all the electrical systems on the car. The Volkswagen Beetle was for many years the exception to this rule on British roads. Not only did the early Beetles have a 6 volt electrical system but also they were wired negative earth. There is no clear reason why any manufacturer adopted one or the other of these systems. It has been argued that one or the other is better from the corrosion point of view, but as far as I am aware there is little or no supporting evidence.

What is a clear technical advantage is if everyone adopts the same standard, and here we have had to fall into line with other countries. The standard is now negative earth, and it looks as if that is fixed for all time. In past years the earth polarity of a car had little significance other than remembering which way round to put in the battery if it were taken out. Now, however, with the proliferation of polarity-sensitive electronic equipment, it is imperative to ensure that the polarity of the car matches any equipment you may want to install in it.

The usual reason, therefore, for wanting to change round the earth polarity is to fit a modern radio or car alarm, all of which are made to work negative earth – in fact this has been standardised for so long that many pieces of equipment no longer have a polarity warning in their instructions.

Radios made up to the early seventies were often made dual polarity and if you want to fit a radio to a positive earth car it may be worth looking around for one which can be made to fit the car. Sometimes there is an external plug which can be reversed, or there may be soldered connections to turn round inside.

P2, P3, P4, 3 litre Mark I and II and early P6 were positive earth. As it is normally the P4 that I am asked to change, I have listed below all the points that must be attended to on a P4. Other models will differ only in detail.

1. The Battery: The same battery may be used, but the connections to it will have to be reversed. The terminals are normally towards the offside; turn the battery round so that they are towards the centre of the car. The main battery lead will have to have its terminal changed to the larger positive type; if the car already has the clamp type terminals they can be changed over. The main lead should be long enough to reach across the battery tray. If it is not, there will be spare length further forward which can be worked back through the clips on the chassis. The earth lead will need replacing with a longer one, as well as having its terminal changed. It is fastened to a lug on the chassis just behind the battery tray.

2. The ammeter: This has two terminals, and all the wires from one should be transferred to the other, and vice versa. If this is not done it will read in reverse. This part is best done whilst the battery is disconnected.

3. The screenwasher: On a '110' an electric screenwasher was fitted as standard, and many other P4's have had electric washers fitted. The wires to the motor should be reversed. Some screenwasher pumps will pump less efficiently if they rotate backwards; others will actually pump backwards.

4. The coil: The coil as fitted had one terminal marked 'SW' for 'switch' and one marked 'CB' for 'contact breaker'. The connections to this should be reversed. Later coils are marked '+' and '–'; for the negative earth the '–' should be connected to the contact breaker.

5. The dynamo: A dynamo is capable of charging a battery either way round without any wiring alterations. The only factor that determines which way round it will 'wake up' as the engine turns is the small amount of remanent magnetism left in the dynamo yoke from the last time it was energised. I have always found that the safest way to repolarise the dynamo is to first change round the battery, then disconnect the wire from the 'F' terminal on the regulator. Touch this wire five or six times on to the 'A' or 'B' terminal; there will be a small spark produced. Reconnect it to the 'F' terminal – the dynamo is now repolarised. An alternator, incidentally, cannot be changed in this way, as it contains diodes which are set one way round.

6. The radio: If the car already has a radio installed, it must either be changed round or disconnected.

7. The fuel pump: If you purchase a new SU fuel pump, or indeed, if you take one of my rebuilt ones, it will come with a diode to suppress the spark at the contacts instead of the capacitor which was originally used. A diode suppressor will make the contacts last longer, but it then renders the pump polarity sensitive. Any pump with a diode in it should be marked as 'positive earth' or 'negative earth'. If your pump is not marked then you need do nothing with it as it will have a capacitor in it. If it has a diode in it, the diode must be reversed or the pump will not function.

8. Anything else? Look round the car to see if anything else has been fitted which could be polarity sensitive. A built in charger, perhaps, or a charger plug? An alarm? If there is, it must be attended to.

9. Labels: It is a good idea to label the car as negative earth, both by the battery and under the bonnet, for the benefit of anyone else who may be working on it. I was once nearly caught out by a car which was negative earth but not labelled as such.

After all is done, start the engine and check that the charging system is operating correctly. Then fit your new radio!

John Backhouse — ***January 1994***

MORE INSIDE LOCKING ADVICE!

Paul Toth seeks advice on converting his driver's door handle to lock from the inside. As I've just completed the exercise, perhaps a few words may help.

The task could not be simpler, no modifications are necessary, only the removal of a small screw identified from the parts lists as "special drive screw 78111". When the trim is removed – as all the best manuals say, glossing over the problems involved – you find the part of the door latch mechanism which carries the interior handle is held on by three screws. Carefully removing these and moving the mechanism to the right exposes the screw in the face of the latch mechanism. It's quite easy to remove and reassembly (as the manual always says) is in reverse order. The parts list shows that whether a left hand drive or right hand drive car, the locking handle was always on the right hand door so those overseas don't need to modify these instructions!

Of course there is a snag and I've experienced it! We all know that P4's were built with everything organised for a purpose, so the drive screw wasn't there by accident.

This modification allows you to lock yourself in – but it also allows you to lock yourself out! The outer lock doesn't interact so locking the driver's door from the inside and closing it gives absolute security. A bent letter opener and a steel rule should be carried in the boot for use in emergencies for tweaking the quarterlight open – enclose a S.A.E. for sketch! Then you can reach in and open the door from the inside. I will keep the modification as there have been times when I or my wife wished we could lock the driver door – but the entry kit stays in the boot.

Ian P. Harris ***November 1992***

DISTRIBUTING THE SPARK

Many of our P4's are still operating with the original distributor as fitted in the Rover factory on their initial build and, of course, their mileage accumulation may be great or relatively small. If the spindle of the unit has been given a small, but regular shot of lubrication throughout its life, it may well be performing a good job of supplying a steady impetus for the engine's ignition to function satisfactorily.

However, in view of the amount of calls I get concerning various ignition problems, I thought I would jot a few notes down to help clear a few points (no pun intended!).

Firstly, on the mechanical side of the distributor, there are only a few components that are prone to wear, these being: the main spindle and bush (on earlier models, a ball race bearing is incorporated and can develop wear after high mileages, especially with neglect on maintaining lubrication). Slight wear can occur on the top spindle integral with the cam-lobes and balance weight anchors. Obviously, any wear in the aforementioned parts will give way to intermittent opening of the contact points, because the cam-lobes are not revolving true to the axis of the spindle. This fluctuation can usually be seen when a strobe light is used to check the ignition timing. If you make a thorough job of cleaning the timing marks on the flywheel and then carefully paint the 'before-top-dead-centre' timing line that is correct for your particular model with either white enamel or the typist's friend – Tippex – you will then see the marks very clearly with the engine running and the strobe is flashing at the No. 1 firing point. With the engine running at tick-over, the timing mark should stay devoted to the marks. Any fluctuation from either side of the line will show the amount of wear that is present. At this stage, you can increase the revs of the engine and check that the advance/retard mechanism is working correctly, i.e. the timing mark should promptly disappear out of the fly-wheel cover aperture or window. If the line stays stubbornly where it is, it either means that the diaphragm is ruptured in the advance/retard housing, or that the balance weights or springs are not functioning as they should. The diaphragm can be checked by sucking on the end of its chamber after removing the capillary pipe attached to it and observing whether the swivel lever is moving inwards towards the chamber. These are rare beasts to find, but I should think Geoffrey Kent will be able to supply you with an excellent replacement.

The little advance/retard springs can become fatigued and stretched on high mileage P4's but I often find that they have been damaged on a previous overhaul by the use of pliers to unhook them from their posts. The only way to remove them is with a strong squeeze between finger and thumb! That way, you will not distort the spring in any way.

These springs vary from model to model, because their torsion works in conjunction with the balance weights. Again these Lucas springs are hard to locate these days.

By and large, the most common fault is wear in the main spindle and bush. Unless you are an engineer and have access to machine-shop facilities, this is not a job you can tackle yourself as it involves dismantling the distributor completely and making a new bush. When I repair distributors, I make the new bush from phosphor bronze as this has a very high resistance to wear and also, of course, is much more suited to the job than the sintered cast metal used in the original mass produced item. Phosphor bronze would have been prohibitive for a high production run. If you are able to renovate your own distributor, be careful not to make the bush – which has two outside diameters – too tight a fit in the body of the distributor. A maximum of .0005" (.0127mm) interference fit is necessary and even then, the bush should be fitted cold from the freezer and the distributor body hot from being dunked in a pan of boiling water for five minutes. Trying to fit the bush without this preparatory precaution, could result in cracking the unit. Once the bush is fitted, you then carefully hone the bush to give a total working clearance of .0004" (.010mm). After honing very meticulously, clean the bush and spindle before lubricating and re-assembling. Do not forget to drill a new oil hole in the bush!

If you cannot do this rather specialised work yourself, you will either have to look around for a used distributor which has fared rather better than yours, or go to someone like Holden who can supply you with a rebuilt unit.

Really worn distributors can produce quite an awful rattling sound whereby the vibrations can be felt by grasping the distributor cap whilst the engine is running.

If the valves and their seats are in good order along with the ignition leads, plugs and covers and, of course, the carburation and distributor are in tip-top condition, this can really be the icing on the cake for a super smooth P4 engine.

A steady contact breaker gap is important for the ignition coil to do its job to the full too, as the dwell angle (the number of degrees the points are closed during one ignition cycle) determines the length of time available for primary current to build up in the coil, and therefore produce the strongest possible spark. Couple this with a good condenser and you are in business, provided, of course, the contact breaker gap is set correctly along with the ignition timing. As I have said before, the points gap can be set quite adequately using feeler gauges, but if you have got the use of a dwell meter, all the better for spot on settings. The dwell angles for the P4's are 38°± 3° for all six cylinder engines except the '110' which is 35°± 3° and 60°± 3° for the four cylinder '60's and '80's. Do not forget to lubricate the steady post and cam lobes on fitting new points to the distributor.

Another cause for ignition problems which can be annoyingly intermittent, is the break-down of the tiny low tension wire which is attached, but insulated, to the inside of the distributor body feeding through from the low tension wire attaches to the outside of the body. This very important little piece of wire attached, along with the condenser wire, to the moving arm of the points is made up of many very fine wires. The original wire used was cotton covered and over the years literally falls to pieces. If you go into an electronic spares shop like RS or Maplin and explain the type of wire you require, they will be happy to sell you a modern plastic covered equivalent which will make an excellent replacement for this often troublesome piece of wire. If you then carefully remove the tags from the old wire, using a soldering iron, you can make up a new replacement by soldering the tags to the wire ends.

If the contact points appear to burn and become pitted after a relatively short mileage and look 'sooty' around the contacts, this is almost certainly due to a badly functioning condenser. This cannot be tested for faults and should just be replaced with a new one. The surface faces of the contacts are very finely coated with a hard layer of metal to prevent oxidation taking place and frequent cleaning will soon wear through this coating and cause mis-firing. So do not clean them too frequently.

Quite a common problem with the P4 ignition system is the fact that very often rain can enter through the join down the middle of the bonnet – underneath the trim strip – and create corrosion in the plug lead ends and around the spark plugs. Careful cleaning of the lead ends with a fine wire brush should get the sparks at full strength again.

A persistent mis-firing can sometimes be caused by the suppressors fitted between the leads and the spark plug covers. The late '95' and '110' were fitted with suppressed leads in place of separate suppressors and can also break down causing intermittent firing. If your car is fitted with suppressors and you cannot detect an ignition problem, simply remove them from the leads and run the engine without them. If the car runs smooth under all driving conditions, you will then know the culprit.

One last thing to check is the low tension between the CB terminal on the coil to the terminal on the side of the distributor. This lead can sometimes break down, particularly if it hasn't been replaced over many years of service.

Sometimes a 'cracking' sound can be heard when the engine is pulling hard. This is usually due to a split in one or more of the spark plug covers, causing the spark to arc across to the cylinder head. The best check for this condition, is to run the engine in a dark garage whereby you might well witness quite a 'firework' display!

If, after cleaning and gapping the spark plugs – or better still replacing them – your engine is still running unevenly, you will then have to forget the ignition and carry out a compression test on each cylinder to determine whether you have burnt out a valve. More about that in a future article.

Stan Johnstone ***March 1994***

SWIVEL PINS, BUSHES AND COTTER-PINS

You will have read in Matt's editorial, this edition, that I had experienced problems with his swivel pin bushes and cotter-pins.

Matt had spotted a full set (both sides) of swivel pins, bushes, ball-race, seals and cotter-pins on the auto-jumble stand at NEC and asked me to inspect them before he bought them.

I went along to the stand to have a look and, to be quite honest, was only concerned that the swivel pins were correct and properly ground with the right number of splines to suit the top-anchor. Some were made a few years ago with one spline too many would you believe?

All the parts were in Leyland boxes and the pins were in first class condition. I didn't even think to look at the rest of the parts. Silly me!

Once Matt had given me both radius arms and stub axles, I got to work on replacing the pins and bushes. To my horror, I opened the bags with the bushes in. I couldn't believe my eyes. The bushes had been so-called turned by someone I can only describe as an animal. They had been squeezed so tightly in the chuck jaws of the lathe, that the seal counterbores were like old three-penny bits. No way would the seals have entered these bores, let alone remained round to seal round the pin.

Not only that, but the outside diameter of the bushes was at least .004 thou/inch oversize making them far too tight for fitting into their housings in the stub-axle.

To top it all, the bores of the bushes were minus .010 to .014 thou/inch which is far too much for a reamer to contend with.

The only saving grace, of course, was the fact that they were in a 'metal-on' condition and I was able to re-machine them prior to fitting.

That done, I proceeded with the build-up after reaming the bushes in-situ using the Guild's adjustable reamers. With the pin fitting nicely, I started to fit the cotter-pins only to discover that they were of the later 5/16" thread type (which is no problem) but after a bit of a struggle, I measured them to find that they were 10mm diameter and not the 3/8" (.375 inch) they should have been.

So the job became a nightmare rather that the pleasant task it should have been. So beware – there are some cowboys about! They caught me out, so be careful when buying from NON-ROVER partsmen at auto-jumbles.

The only good thing about the whole affair was the fact that I didn't get Matt's '90' ready for the Norwich Union run and had to use my '95'. I did enjoy myself going round Silverstone, even though it cost me about three months tyre wear in one day!

Stan Johnstone ***July 1994***

DID YOU KNOW…?

…That all is not lost if you lose the number of the key which fits boot and glovebox on a P4. On the '95' and '110' the number is stamped on the plate which can be seen by opening the passenger front door, but on earlier models it is stamped on the top of the radiator. The number is on the left hand end of the radiator, and should be read by standing to the side of the car. There will be three digits, roughly 3/8" high. The two letters which precede the number (FP, FS, FK, FR) can be found by trying other keys in the lock, as the letters define the shape of the flutes cut into the side of the key. There may also be a check here with the numbers, as not every number combination was available with each letter group. There was a time when the local accessory shop would have several cards on the wall and could select a key when quoted the number. Now, however, these cards are held by the specialist suppliers who have to be approached either by post or at autojumbles.

… That on all P4's except the Cyclops the petrol pick-up pipe is well over to the right hand side of the petrol tank. This little snippet may be of no great interest, you may think, but it is of significance if you have already covered some miles switched on to reserve and are desperately looking for a filling station. The general advice is to stay on the crown of the road as much as possible, for if you stay on a nearside camber for too long the pump will suck air and the engine will cut out. I once watched someone who ran out of petrol and parked on a nearside camber. He fetched a gallon of petrol in a can and poured it into the tank. The car would still not start until it was pushed into the centre of the road – and a P4 is no light weight!

All the other Rover cars around this period – late P2, P3, Cyclops, P5 – have their pick-up in the centre of the tank and do not suffer the problem. With the late P4 it seems that there was a competition for space with the spare wheel tray which forced the pick-up on to one side, and the right hand side was chosen so that the pump would not have to compete for space with the filler neck.

. . . That on all the P4s which have an oil bath air cleaner, removal is much easier that a lot of people think. The air cleaner has to come off for a number of jobs, including changing the fan belt. The workshop manual instructs to "Unscrew the air cleaner" which does not give much away, and the novice can easily spend a knuckle-scraping hour undoing the long bolt directly under the air cleaner, and the small one bracing it to the dynamo bracket. Once you know, however, removal could hardly be simpler. All you have to do is turn the whole air cleaner anti-clockwise and it will unscrew itself from a bolt welded on to the bracket. On the very early models the bolt is loose and must be held from underneath. Many a time I have removed the air cleaner from a customer's car to hear an exclamation of 'I wish I'd known that a week ago!'

Jon Backhouse ***September 1994***

CARB SWAP FOR A ROVER '80'

We've had VVY 444 for about 5 years now, and although she's got just 33,000 miles on the clock, the old girl did occasionally (well, to be truthful, every time the loud pedal was pressed actually) show signs of breathing difficulty, thanks to Mr Solex and his petroleum spirit vaporising instrument.

444 also had another habit when hot, in that each time the engine was stopped, a float bowl full of 4 star skilfully crept out of its allotted receptacle,

and via the choke tube and throttle spindle, make like a Lemming onto the manifold.

Now, in spite of being a Yorkshireman (with a bit of Dane mixed in) I have not calculated the value of the lost fuel, but did notice the exorbitant price of a replacement starter motor. This was of some relevance, as the excesses of BP's finest that came to rest in the inlet manifold, meant that this almost priceless article, had to work overtime, spinning the flywheel until 2286cc took over. I had thought about leaving the car in gear during this procedure in order that some of the time could be spent travelling.

So, the threat of needless expense, along with the soot-fall from the exhaust when she did fire up (which incidentally, I found slightly embarrassing, as folk were beginning to confuse my car with Wheater's) prompted some investigations.

Overhauling the Solex failed to solve the problem of flooding when hot, which was my main concern, as to be fair, 444 always started instantly when cold and managed 30 - 32 mpg with ease. The flat spots, which come as standard with the Solex, at light throttle openings, I could live with.

Back issues of Overdrive suggested various alternative carbs., but I could find no mention of a fairly obvious replacement – a Zenith 361V.

As most of you will know, the P4 '80' shares the same underbonnet accessory as that of the farmer's favourite, the Land Rover. A few enquiries soon revealed that the Agriculture impecunious had also got cheesed off with the Solex, so Land Rover had recommended the Zenith as the alternative.

A trip to the local 4WD shop led to a brand new carb and cast adaptor plate, which allows the Zenith to be turned through 90°, finding its way into my hands. A bendy friend helped to take the sting out of the bill for £98, and off we went.

Closer examination revealed only one small hitch in the fitting process. Solex spindle rotates clockwise, Zenith anti-clockwise. It soon became apparent that the simple expedient of turning the carb pivot shaft through 180°, and fitting the control rod at the other side of the shaft wouldn't work. The subsequent ministrations of a hacksaw and welding torch resulted in the lug holding the pivot ball being transplanted to the other side of the shaft – success! Everything fitted and turned as it should. Pipe work was replaced, which left only the choke cable to sort out. The original was about 4" too short as it now had to bend through 90° to meet the mechanism. Wanting to retain the original outer cable I pondered and rummaged, and found a spare valve lifter cable for my 1958 Panther motorbike. This had the correct nipple already soldered on one end, and fitted straight into the cold start knob. A short extra length of outer cable completed the job.

The final task involved a bit of 3 dimensional thought – the overdrive micro switches. Remove the cam from the end of the shaft, turn it round and then rotate it through 90°. Then swap over the pairs of wires in the switches, adjust them as outlined in the manual and away you go.

First impressions are of a smoother tickover, less intake noise, and an engine that pulls much more smoothly at low revs.

A run down to the National revealed no problems, and a steady 60 - 70 mph cruise along mixed roads and two city centres showed a consumption approaching 35 mpg. In fact I only stopped at a garage once on the 350 miles and I wasn't sure whether to put more petrol in or to drain some out.

Should anyone out there want to try this and needs further details please contact me. I can do the mod to the carb shaft, free of charge if required.

Oh, and no she doesn't flood any more.

Steve Shone ***July 1995***

DUKE OF HAZARD

Whilst continuing the overhaul of my P4 '100' over the winter months, I came across two spare insulated bullet socket connectors attached to two leads emanating from the main cable harness, inside the instruments compartment.

What use could they be, they obviously had been fitted for a purpose.

This interested me and I traced them to the 2 - LH and 2 - RH winker lamps.

I like to think that the Rover engineers, with characteristic foresight, had way back in time thought about Hazard Warning Lights, and so had the harnesses manufactured with these unused bullet connector sockets added in anticipation of such future use.

With Rover having done the difficult part of the wiring, it would be easy to fit a Hazard Warning Light System, controlled from a Single Pole switch:-

'Up' for Normal Winker operation, 'Down' for Hazard operation.

The essential requirements are that:

1. The original wiring is not disturbed in any way except for the insertion of the single pole, Hazard Warning On/Off switch, which is tucked away out of sight under the dash, and connections to the Winkers made via the aforementioned bullet connectors.

2. The Hazard Warning Lights must be able to be switched ON with the car electrics in any operational position, i.e. with the ignition switched off etc.

3. The main wiring must be safe (this has already been done by the Rover Co.).

4. The Hazard Circuit must be completely isolated from the Normal Winker operation Circuit.

5. The Default condition, from any untoward situation, must always be an immediate return to Normal Winker working by returning the Hazard switch to the UP (off) position.

These conditions are achieved by using the following components.

1 - Single pole switch.
1 - Hazard Flasher Unit type. 9 FL12V with holder.
1 - 12 volt D.C. 3PDT (3 pole double throw) relay.
1 - P600A Diode.
1 - 12 volt lamp (red) with holder.
Cost approx. £10 to £12 from Maplin Electronics.

I found the ideal position for the Red Hazard Warning Indicator lamp, was to fit it out of sight, just above the instrument panel cluster. When ON the light from the lamp reflects strongly from the chrome rims of the instruments which is quite sufficient to show that the Hazard Warning Switch is ON.

Don't forget to disconnect the battery under the rear seat when working behind the instrument panel, or better still, consider fitting a battery isolator switch to the bulkhead near the starter relay, which is an ideal place, just waiting for such a switch.

I wonder if all P4 cars have a cable harness with the two spare winker bullet sockets fitted, like the harness fitted to my P4 '100'.

If not, one can always locate the winker leads source from the main junction box on the bulkhead near the steering column.

If you think that my experience will be of interest or help to anyone, I will be pleased to give you a short article with fitting instructions, together with a simple circuit diagram.

Richard Anderson — ***July 1995***

SPOT THE DIFFERENCE

Yes, a British Jaeger with a sweep secondhand. How is this, you may ask!!

Well we all know the problems with these little devils, stopping now and then for no apparent reason. Give them a kickstart and away they go till the next time. What with losing in the summer and gaining in the winter. This incidentally is due to the viscosity of the oil, (lubricating the mechanism) varying and causing the balance wheel to either long or short stroke. Over the years I have cleaned movements, repaired balance wheels with contact pins, replaced strikers and coils.

Well just by chance one day I was browsing in one of these cheapy arcade shops and I came across some all polastic quartz travel alarm clocks for £2.49 each. Having checked the accuracy of the movement I then stripped out the quartz mechanism to see if it could be accommodated in the case of the Jaeger, and yes with a few modifications it would.

So with a junior hacksaw, file, soldering iron, solder and wire I removed all electrical and mechanical components associated with the alarm mechanism. I then cut the circuit board away to fit into the Jaeger case. Where I had to cut away the copper track, this was hard wired back.

Next was to mount the quartz movement and backface (the part with British Jaeger printed on it) to the correct hands to back face relationship as in the Jaeger.

This was achieved by marking off the back onto 1mm thick insulation board, drilling the centre hole and then piercing the 3 face fixing holes just large enough for the original screws to cut their own threads.

Next was to drill four holes in the insulation board, two for two long 4ba screws, to fix the quartz movement to and finally assemble the whole thing in the Jaeger case, one for a small screw to accept the original hand set/start button and one for the fast/slow segmented wheel indicator. Next is to cut away as much of the insulation board as possible, but leaving a small three point mounting, this will centralise the movement and allow a light path to illuminate the front face. Next is to drill the circuit board and assemble the backface and quartz movement. The hands are the original and unless you are lucky, will need to be either opened up or soldered up and redrilled. The centre sweep has the original, though modified, black button fixed to it. Should you not require this facility thus giving more originality, cut off the sweephand and just the black button will rotate virtually unnoticeable.

The back of the Jaeger case can now be drilled to accept the 4ba screws and a hole for the hand set adjustment, Now solder neg and pos wires to the circuit board for the 1.5v AA size battery. Now blank off all holes not required to stop light escapement, and assemble and place in the Jaeger case. ***Note* The 4ba fixing screws** must pass through **grommeted holes** as should the **battery wires** and be very careful when assembling that none of the quartz movement can short out to the Jaeger case, as this will be at a 12v potential with the lamp illuminating the clock. Now make a small clip to hold the battery to the back of the clock, this and the battery **must be insulated** from the case. Now you have as near original looking British Jaeger clock that keeps good time with a bonus of telling you the time when the battery is disconnected. Not bad for £2.49 plus a week of evening work. Better and more interesting that the telly!!

P.S.

Some of you may have the later pattern back face with an extended 45° bezel, either cut it away or be very careful that nothing shorts between case and quartz movement. “Best of Luck”.

Mike Coleman ***January 1996***

P4 CYLINDER HEADS

The top-end overhaul of the P4 is well documented and even the workshop manual is quite useful if followed carefully. However, there are a few failings that are not covered in the manual and it is these that I would like to highlight in the hope that they will be of use when the head is removed for inspection.

The cylinder head is quite straightforward to remove but generally speaking, the only likely snag is the removal of the temperature transmitter from the outside of the head (not '95', '110'). The brass face of the bulb has usually been overtightened to the taper seat in the head and being dissimilar metals corrode together. They should be assembled with vaseline or copperslip on the seat and threads but rarely are. The capillary tube attached to the bulb is very delicate and care must be taken not to bend it as this will surely fracture it and it will become useless. If this does happen the whole unit from dial to bulb will have to be replaced.

Care must also be taken with the oil-feed pipe into the rear of the head. It should be undone using a well fitting ring spanner and try not to bend the pipe in any way, otherwise it will be difficult to restart the thread on re-assembly. Always drain the block as well as the radiator before commencing work, otherwise you will find some of the cylinders full of coolant on removal.

Once removed, the head should first be inspected for excessive corrosion in the water-ways. By laying the old cylinder head gasket on top of the face you will be able to see how far any corrosion has eroded beyond the cut-outs of the gasket. For example, if it has travelled too close to the combustion chamber, a bolt hole or too near the edge of the head. If the corrosion is excessive the easiest way out is to try and locate a head that is in better condition. These secondhand heads are not usually too expensive to buy, but the problem of course, is finding one. The other option is to find an engineering firm that is able and willing to machine out the corroded pockets and then fill with aluminium weld. This is a highly specialised job and could well be very expensive, but once welded the water-ways can be machined out and the head re-faced.

Refacing must be carried out correctly, otherwise it is better not to machine it at all. Most engineering firms will fly-cut the head on a milling machine. This is not good practice as the surface finish is not fine enough. Rover always specified that the heads were ground on a surface grinder (not plough-grinder) and then polished in a straight line using 1000 wet-or-dry paper adhered to a flat plate at least the width of the head and lubricated with paraffin. The heads should not be clamped down for machining but should be bolted to two flat plates (steel) about 4½" x 6" x ¾" using ⁵⁄₁₆" bolts through two of the valve guides. The plates should have a ⁵⁄₁₆" tapped hole in the centre of them for the bolts to screw into. These plates will then hold on to the magnetic chuck of the surface grinder. ('110' heads will require a shallow counter-bore in the plates for the protruding guide to sit into)

Nine out of ten heads I refurbish are warped to some degree. At best they will need .005" removed to clean them, but I have known them to need .010". Incidentally, on the '110' head the step-down to the boss for the waterpump to meet the O seal should measure .125" (3.175mm). If too much is removed from the cylinder head face, the gap could become too small and over compress the seal.

Invariably the by-pass holes in the head between the pistons are clogged up. These must be carefully drilled out, using a hand-drill with 3⁄16" drill bit. The position of these can be ascertained by using the gasket as a template.

When stripping the head prior to machining, always number the rockers before removal. Once removed, the rocker pads should be examined for any indentations of the valve stem, and for accurate tappet settings should be stoned smooth by pressing down on an oiled oil-stone in a to-and-fro action following the radius of the face.

Take note also of the rocker spacers (not '110' or '80') these can wear to such an extent that the rocker face is not central with the valve stem and can lead to noisy inlet valve clatter. They are not expensive and are well worth replacing.

Check that the by-pass hole at the base of the thermostat aperture is clear, if not, drill out using a hand-drill and a ¼" drill bit. This is not easy to check on the '110' as it is drilled at an angle, an air-line is useful here.

Of course, before any machining is undertaken it is an ideal opportunity to have the head professionally cleaned. Try to avoid grit or bead blasting because this can leave nasty abrasives hidden in the many crevices of the complex casting. The best method of cleaning is vapour-blasting which does not leave any residue and gives a good satin finish.

While the head is off, clean the spark plug threads, ideally with a 14mm tap or an old plug with 3 or 4 slots cut longitudinally into the thread.

Before fitting the head, make sure that the threads in the engine block are thoroughly clean and free of any oil that may be lurking down the bottom of the holes. If this is not done the cylinder head bolt could force any trapped oil to crack the block, because oil cannot be compressed.

When the head is built and prior to fitting, smear its face with an even coating of clean engine oil, also the engine block. Do not use grease, hylomar or hermetite. Torque the head down in increments of 10 ft lbs until the required torque is met. Always use a good quality torque-wrench and never guess!

Never use a wax-stat type thermostat, it must be of the bellows type and use a strong solution of antifreeze on re-filling and do not forget to bleed the system when the engine is at full temperature.

Stan Johnstone ***January 1997***

AVON TYRES RECOMMENDATIONS FOR TYRE STORAGE

Introduction
In view of the harmful influence of temperature, humidity and light, inside storage is essential.

Humidity
The store room should be cool, dry and moderately ventilated. Moist conditions should be avoided. Care must be taken to ensure no condensation occurs.
Tyres destined for retreading/repairing must be thoroughly dried out beforehand.

Light
There should be protection from sunlight and strong artificial light with a high ultra-violet content. Room lighting with ordinary incandescent lamps is preferable to fluorescent tubes.

Temperature
The storage temperature should be below 25°C and preferably below 15°C. At temperatures exceeding 25°C certain forms of deterioration may be accelerated sufficiently to affect the ultimate service life. Direct contact with pipes and radiators must be avoided.
The effects of low temperature are not permanently deleterious, but can cause the products to stiffen. Care should be taken therefore to avoid distorting them during handling at that temperature. When they are taken from low temperature storage for immediate use, their temperature should be raised to approximately 20°C throughout before they are put into service.

Oxygen

As ozone is particularly harmful, storage rooms should not contain any equipment generating ozone such as fluorescent lighting, mercury vapour lamps, electric motors or other electrical discharges. Combustion gases and vapours which may produce ozone via photochemical processes should also be excluded.

Solvents, fuels, lubricants, chemicals, acids, disinfectants and the like should not be kept in the store rooms. Rubber solutions should be stored in a separate room and the administrative regulations on the storage and handling of inflammable liquids must be observed.

Deformation

Products should be stored in a relaxed condition free from tension, compression or other deformation since these may cause cracking or permanent distortion.

Rotation of stock

To avoid deterioration storage time must be minimised. Stocks should be issued from the stores in rotation so that those remaining in storage are of the latest manufacture or delivery.

Short term storage

For short term storage (up to 4 weeks) tyres can be stacked horizontally, one on top of another, on wooden gratings but the height of the stacks should not exceed 1.20 metres. After 4 weeks the tyres should be restacked, reversing the order of the tyres.

Long term storage.

For long term storage tyres should be stored upright in a single layer on shelf racks with at least 10cm clearance above the floor. To avoid deformation it is advisable to rotate them slightly once a month.

Tubes

Tyre tubes should either be slightly inflated, dusted with talcum and placed in the tyres, or stored in a deflated condition in small stacks – maximum height 50cm – in the compartments of shelf racks with a level bottom. Slatted pallettes are not suitable since they might apply pressure at particular points.

If tubes are supplied by the manufacturers in cartons or wrapped in film, they should be left in these because the packing provides some degree of protection against contamination, oxygen and the effect of light.

Flaps
Flaps should preferably be placed with the tubes, inside tyres, but if stored separately they should be laid flat on shelves free from contamination from dust, grease and moisture. Never suspend them – this can cause deformation and elongation.

For further information please contact: Avon Tyres Ltd., Melksham, Wiltshire SN12 8AA. Telephone 01225 703101. Telex: 44142. Fax: 01225 707880.

ROVER P4 SCUTTLE AND A-POST REPAIR TECHNIQUES

Barry Kensett of Yorkshire has produced this excellent practical article on welding on the P4, including the difficult A-post area.

In the late fifties and sixties, the corrosion resistance designed and built into our cars was far inferior to current day practice. The Rover may be considered to be better built than most but some of the design features are appalling to today's standards. The materials and techniques available were also inferior in most respects, it was unusual to use zinc coated materials (unless you had a Volvo) which were difficult to spot weld and paint and were more costly than bright steel. There were no effective interlay compounds which could be spot-welded through to give protection in the many flange joints, there were many closed boxes in the body structure which had no internal paint or wax treatments and little thought was given to drainage leading to water traps.

The area behind the front wheels of the P4 is very prone to corrosion and it is extremely rare to find a car that has not been damaged in this area. Figure 1 illustrates the components likely to be affected to some degree.

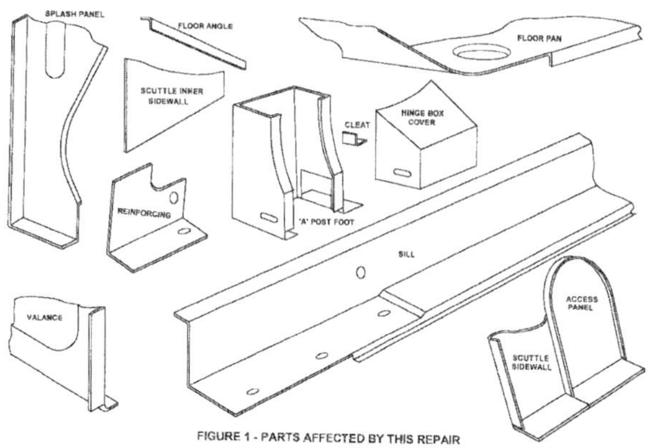

Figure 1. Parts affected by this repair.

It is not difficult to review the phases of the corrosion process which starts with the effective blasting of the splash panel and valance behind the front wheels with sand, water and saline solution. The rubber flap seal on the splash panel may not be a good seal. Once the splash panel is penetrated, the saline solution is partially trapped in the closed box which did not originally have the benefit of any drain holes. The bottom of the sill extension, the scuttle sidewall and access panel (this exists to let the spot-welder arm in manufacture) and the parts around the base of the A-post are immersed and quickly corrode, particularly where the many flanged and spot welded joints provide electrolytic batteries. Keeping the car in a nice warm garage would accelerate the process! Rain from the top adds to the problem as the only escape path for water draining from the windscreen and bonnet area is down the outside of the scuttle and onto the A-post from above. It is possible to see into the base of the A-post through the access hole behind the scuttle trim panel but this only reveals part of the problem area.

Meanwhile a second attack would be taking place once the inner valance is penetrated allowing water into the trap formed by the horizontal spot-welded joint between the scuttle inner sidewall and the vertical face of the sill extension. A path is now open for the sand-blasting to strike the floor in the footwell area around the handbrake apertures and perforation here usually occurs. Once the box around the lower door hinge is perforated, water enters this area affecting the door hinge itself and then running back under the sill mats to perforate the sill from above. Eventually the sill structure around the body mount starts to break out compounded by the breakdown of the outer end of the front outrigger which is another repair topic in its own right.

There is no standard repair for the areas of damage, the rate of corrosion depends on the usage of the car. It will be different on both sides of the car and the course of events will be modified by any superficial repairs which may have been carried out during the car's life.

On my own car I knew that the usual problems existed but were masked by a plate which had been welded to the bottom of the front wing and folded up and tacked to the scuttle sidewall. This was obviously done some years ago as it had perforated in a couple of places. On the other side it was possible to hear the rust debris bouncing around if the bottom was tapped with a hand! The timing of the repair was eventually precipitated by the breakdown of the body mounting in the sill and the outrigger and I decided to tackle the whole area as a winter project.

The Repair Process

It is necessary to remove the front wings as described in the workshop manual; in my case it was necessary to cut the wing at the base due to the temporary repair. The front valance had also been tack welded to the wing! The wing must come off before the doors to get access to the bolt through the A-post; the hinge bolt through the sill will probably need grinding off. The doors and hinges can then be removed and care must be taken with the sponge seals unless they are due for replacement. I would recommend that the bonnet is also removed. Front seats, carpets and handbrake will need to come out if the floor is perforated and the aluminium trim strip along the top of the sill will need removal followed by any insulation pads within welding range. The front screws of this trim may need the gentle application of an impact driver to loosen them. Underseal will need to be removed in areas where welding is needed; a hot air gun and scraper is the easiest way to do this. The rubber seal on the splash panel should be removed if it hasn't dropped off.

Offside 1 – Before

The car must be raised and if a lift or pit is not available, I find that lifting the car about two feet is about the right compromise between access and arm reach. I removed the front outriggers as these were also due for replacement and access to the sill is easier without these.

Some repair parts are available, I found that the sill extension section and the scuttle side access panel save some manufacturing time. The section available for the scuttle sidewall was a quarter of an inch too narrow for my car; I made a piece to fit. I used 18 swg mild steel sheet for all repairs and I made patterns with card first before cutting the metal. I use a nibbler to roughly shape the sheet and Gilbows and file to finish off. All folding for this job can be carried out in the vice with a few bars and a hard wood block. Before attacking with a saw and grinder, the job can do with a lot of hard looking at to determine the best place for joining to sound metal. The old joiner's adage "keep your wood as long as you can as long as you can" applies. Care must be taken in not cutting away too much at one go and risk losing the shape.

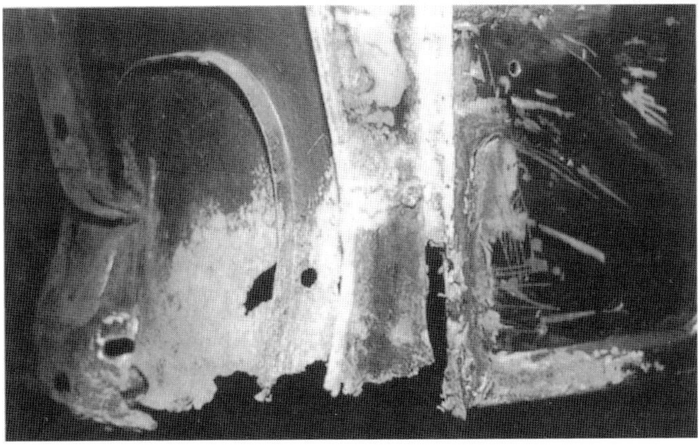

Offside 2 – Before

The scuttle sidewall should be cut back to sound metal and the access panel ground out to help access to the sidewall.

The sill repair section is sized to lap over rather than butt weld to the sill; this means that any sound sill can be left in place which assists with alignment. If there is damage to the sill behind the coverage of the repair piece this should be repaired with inserts, butt welded in place. Due to the corrosion trap at the top flange of the front end of the sill, it is unlikely that there is sufficient material hanging from the scuttle inner wall to meet the new section. This must either be replaced by a lap welded extension or plug welded to the angle on the floor although the latter is tricky due to the thickness of metal on the down hanging flange and the difficult access. The sill repair can be clamped in place at the body mount position and lined up with the remains of the splash panel. When confident on position, the holes for the body mount hinge bolt can be drilled to allow the section to be bolted up.

An A-post foot can then be repaired as necessary; on one side I had to replace the front wall, back and base. The hinge box cover can then be welded around the outside and the hinge should be offered up at this stage to make sure that there is adequate clearance.

The floor pan should be repaired before the sill is welded in for better access. I used insert repairs but lap welds would be satisfactory here if you do not mind the appearance. Shaping the pressed area around the handbrake hole is a bit tricky; I beat it roughly to shape on the bench, welded it in and then heated the double curvature to bright red and tapped into shape in situ.

The sill extension can then be attached permanently and for this I found that bronze welding is best as it penetrates the lap better than welding. This is attached at the rear end of the horizontal and vertical laps from below the car and to the A-post foot faces from the top. The scuttle inner wall is then connected to the vertical flange and I used rivets here to enable the use of a chromate interlay material to prevent corrosion at this vulnerable joint. Plug welding would be satisfactory and as good as the original build.

The scuttle access panel can now be offered up and the forward section of sidewall made to suit. This is butt welded to the sidewall and the access panel is edge welded in place. The aft flange is heated and tapped back close to the A-post and welded. The bottom flanges are bronze welded to the sill section (remove the zinc coating first).

The splash panel can now be cut back to sound metal and a new section fabricated and butt welded in place. Similarly the valance needs an insert repair and finally the four flanges of the valance, sill, sidewall and splash panel are edge welded. I sealed all joints in this area with bronze weld to help weather resistance and close up holes exposed to the front wheels.

If the splash panel is badly corroded higher up, it is possible to obtain a new repair panel. Mine was not bad enough for this but was perforated intermittently adjacent to the forward projecting flange. Welding in this area would not be successful, there is insufficient sound metal to go at and the heat would set fire to the trim behind the dash. I straightened and ground the flange to the original projection. I then made a section of angle to fit onto the face of the splash panel far enough away to be lap welded and flanged up and over the projecting flange and clamped with chromate interlay rather like a Mini body seam.

All metal was then primed with a damp resistant primer (e.g. Rustoleum 769), seams sealed with 3M body caulk, black enamelled and waxoyl black underseal to finish.

Bottom door hinges should be checked for wear and repaired if necessary (see Overdrive No. 95, March 1994, Page 20) before replacing doors. Front wings and bonnet are replaced as described in workshop manual. With care to rustproofing, the above repair job should never need to be done again.

Offside 1 – After

Offside 2 – After

Floor pan repair

ROVER P4 FRONT OUTRIGGER REPLACEMENT

The front outriggers on the Rover chassis are subjected to the blast of water and salt from the front wheels and not surprisingly are a common corrosion point. The forward face and the bottom section usually suffer first and these can be patched. When the outer face which supports the body mounting starts to break up there is little access to do a permanent repair and a more drastic treatment is required. An additional major problem is caused by the collection of water in the bottom of the outrigger which cannot escape but can penetrate the joint with the main chassis member and corrode through the chassis at this point.

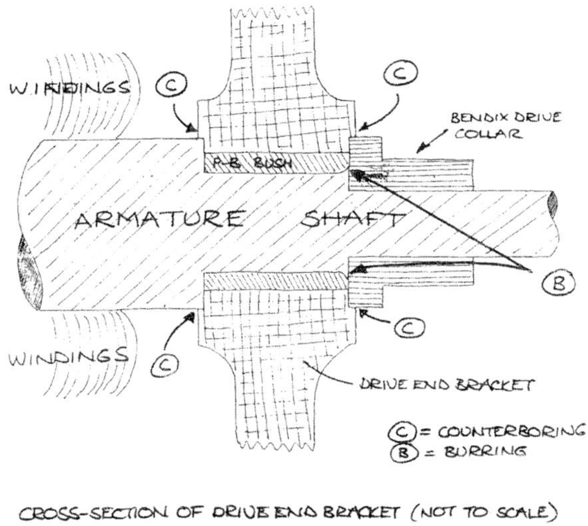

This diagram first appeared in Freewheel, the bulletin of the R.S.R.

The best access to carry out replacement is with the body off; I have heard of people who raise the body about a foot which can usually be done in the garage with normal lifting tackle. I have also heard of people waving their welding torch blind in the gap between the outrigger and the floor – the mind boggles!

Access in floor to remove and replace top flange and outrigger.

Faced with the breakout of the body mounting I talked to a few experts and decided to change the outriggers without removing the body, and to get access to the top I cut two access holes in the floor nine inches by three inches using a nobble (had I used a grinder there was a danger of nicking the hydraulic pipes on the offside). I then jacked the car up about two feet and removed the master cylinder and its rod along with the handbrake and its front rod. It is then possible to cut the outriggers off with an angle grinder and clean up the faces to receive replacements. There is a slight complication on the offside with the master cylinder mounting bracket; I cut this by gouging into the outrigger to leave the bracket intact for welding to the new outrigger. I needn't have bothered as the new outrigger sat in a slightly different position and the bracket had to be extended anyway. There is no need to take the bracket off of the chassis member.

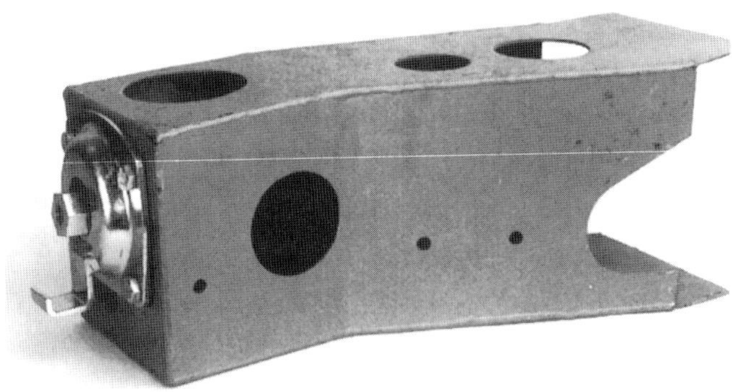

Rearside outrigger.

Once the outriggers were off I found that the chassis was rusted through where water had collected in the bottom flange but this damage had not extended beyond the outrigger flange. I therefore was able to weld an insert into the chassis before fitting the new outriggers.

I elected to use the ready made new outriggers which are on the market and which fitted reasonably well. The flanges were long but were easily cut back. The height was about an eighth of an inch too small to fit over my chassis but I cut the weld for about one inch on the bottom flange which enabled it to be sprung open sufficiently. The cut was then welded up. I had to trim the vertical faces very slightly to get a close fit.

The outer end of the outrigger was located in the sill using new mountings but I used a bolt with a turned down shank to set the outrigger 0.1" high to ensure that it carried its share of body weight. The zinc plate was removed from all welding faces.

Whilst I usually prefer gas welding, I used MIG to attach the outrigger to the chassis as the amount of heat needed would have set fire to the underseal on the floor. The welding is very straightforward but needs at least 130 amps to get the right penetration (this is about the limit on a 13 amp plug). Bronze welding would be strong enough and use less heat but I felt that the slight pitting on the chassis would make it difficult to tin properly, particularly on the overhead weld where balls of molten metal up the sleeve are a hazard at the best of times. I should have painted the inside of the outriggers before welding them on; it would have been a lot easier on the bench!

As mentioned earlier, I extended the master cylinder bracket to meet the outrigger face which sat a little forward of the original. I lapped the extension onto the old bracket and fillet welded it to the outrigger being careful to leave space for the jacking point.

The undersize bolt was removed from the body mounting by removing the bolts through the outrigger that provide clearance to extract it. The sill was then jacked to line up the bolt hole and a packer was made to fill the gap between the mount and the sill. The jacking points, handbrake, master cylinder and rods were then replaced. I finally made up a pair of panels to close up the floor.

Rearside outrigger.

Barry Kensett

WATER ON P4 SPARK PLUGS

I was interested to read Michael Walkling's letter in the September issue of Overdrive as my '110' occasionally suffered from the same problem – wet spark plugs as a result of water dripping through the clips holding down the central chrome strip on top of the bonnet. I can recommend the following solution.

Expanded foam pipe lagging comes in 1 metre lengths and is pre-slit so that it can be clipped around pipes. There is a large flange running under the centre line of the bonnet through which the chrome strip clips pass and hence the water. The pipe lagging can be pushed onto this flange. Any water that enters will run down the central hole of the pipe lagging and will drain down by the radiator. Because it is so light it does not require any glue or other fixing, and it actually looks quite neat. It takes about 30 seconds to fit so those who are worried about corrosion due to lack of air circulation or the possibility of it coming off on a long journey can remove it and re fit it when the car is going to be parked in wet conditions. The foam is heat resistant and flame retardant, so even if it does come off no real damage should result. You will need just under a metre, and the lagging for 15mm pipes which has a total diameter of about 40mm is ideal. My local DIY store sells a pack of 5 x 1 metre lengths for £2.50. I fitted this to my car in October and have had no problems since.

Hugh Richards ***March 1998***

FRONT SUSPENSION OVERHAUL –

Alternative Methods and Renovation Tips

"Eight inches from yer balls, is that all you've got?" exclaimed the missus! I looked up from the November issue of Overdrive, gobsmacked, only to notice that she had been reading the notes I had made of the project I was to undertake during the Christmas break.

Yes eight inches ground clearance from the underside of the radius arm balls, this relates to the workshop manuals normal ride height into which you are expected to tighten up all the relevant nuts and bolts after the overhaul of any part of the suspension. Unfortunately my garage is not blessed with a pit, and even if it were, where on earth does one obtain those special half moon wheels to gain access to the top link pin nuts? Perhaps Stan-Stan the Rover Man may have a set tucked away!!

It was with this in mind, and the fact that a good many of us P4 owners are akinned to our beloved vehicles, i.e. - past our prime, have difficulty negotiating within confined spaces, round and cuddly and built for **comfort** not **speed**. So to this end the following technique was devised:

1. Check that the tyre pressures are correct.
2. Give the vehicle a settling bounce.
3. Measure and **note** the distances between the tops of the front wheel arches and their respective hub centres.
4. Place a spirit level across the front overriders or numberplate box and apply a piece of tape either side of the bubble. And if you have a memory like mine, a piece of tape one side of the level and on its adjacent overrider, so it goes back the right way round, saves a lot of frustration later on.

You can now raise the whole front of the vehicle to give a comfortable working height at the rear of the radius arms. I prefer to have three wheels at a time supporting the vehicle, using wooden blocks under the wheel of the opposite side to which I am working, and supporting the working side at the front chassis end.

Level off the vehicle to the marked spirit level and don't forget to take as many safety precautions as possible in stabilizing the vehicle, **remember you will be underneath it**!!

I usually chock the rear wheels, have a support in close proximity about 1/16" at the junction of the front outrigger - main chassis - gearbox crossmember and not relying solely on the handbrake, wind in the master cylinders to foot pedal connecting rod tight, this gives three wheels locked hydraulically.

'Over the top' you may say, but I live to write this article!!

Now for the removal of the springs. Forget the trolley jack, chain and Rover Part No. 262774. Obtain from your local DIY Store 4 lengths of 5/16" whit or M8 studding (these are usually supplied in 12" lengths) 8 full nuts, 4 half nuts and 4 washers.

Disconnect the torsion bar and shock absorber from the spring support plate. Loosen all 10 bolts fixing the bottom link arms to the spring support plate, just enough to slide in the half nuts. Now remove one of the two bolts closest to the centre of the vehicle. Spin a full nut followed by a washer about 1" down a piece of studding, insert a half nut in the gap and screw up the studding so as it projects enough to accept a full nut on top and tighten up. Now spin bottom full nut and washer up finger tight to support plate. Repeat this operation for the fourth bolt along and then both operations for the other side of the support plate. You will note that the four studding are directly below the bolts for the top tie straps. Remove remaining six bolts and you are now ready to wind down the spring. Apply some oil to the studs to help things along and wind down the nuts evenly. To speed things up a little, I used a box spanner like that in the Rover tool tray and a short series ring spanner on the other end, so as it rotates between the studdings.

Although this is not a fast method, about ten minutes, it does in my opinion give a degree of safety to the DIY mechanic. Now dismantle the rest of the bottom link assembly.

Thoroughly inspect all components for corrosion, the support plates are prone to this and also the coned bump stop both rot badly as there is no drain facility. Support plates should be replaced, bump stops can be reclaimed by prefabricating a new cone and welding on. I had to do one of mine.

Check the bush housings for corrosion as any amount will drastically reduce the life of the new bushes. These should be polished smooth with emery cloth. A piece of ⅜" dia. rod with a slot cut up it, to take the emery cloth, and fitted in an electric or battery drill works well or an engineers bearing scraper and emery cloth to finish.

Before fitting back together, the cone bump stop should have two 5/16" holes drilled at 180° as close to the bottom as possible and aligned front to back on assembly. The spring support plate should also be drilled with a 5/16" drain hole at its lowest point, that is facing out towards the wheel. Reassembly of the bottom link arms can now take place. Use a liberal coating of red rubber grease on the bushes, pins and housings. Leave all four nuts loose by a couple of turns on the link pins. *Note: do not fit spring, bump stop and support plate at this stage.*

Now for the radius arm ball. With a jack about twelve inches away from the radius arm housing, firmly support, but not enough so as to lift the vehicle, the opposite sides radius arm. Remove the five bolts and lower the bottom half of the housing. Lower radius arm and ball, clean up as required, replace ball with red rubber grease, bottom half of housing using only the four vertical bolts and leave them loose by a couple of turns.

The top link arm is next. Before dismantling, tie back the top anchor bush housing to the shock absorber mounting bracket to take any strain off the brake hose. Clean bush housings as previously mentioned. Now underneath the inboard bush housing attached to the chassis, you will find a hole about ¾" dia. in the chassis, also one about ⅜" dia. when viewed from the engine compartment. These are the points of ingress that fills up the bump stops, they will need to be sealed off.

Now re-assemble with red rubber grease leaving the nuts on the pins loose as before. *Note: do not fit the top link bump stop at this stage.*

Now raise the hub assembly to the previously measured distance between wheel arch and hub centre, and pack in position between the top link arm and the rebound plate on the chassis. This is your **ride height.** You can now tighten **all** nuts and bolts with ease. Remove the jack from under the radius arm. If you are going to do the opposite side of the vehicle without driving the car, there is no need at this stage to fit the horizontal bolt on the radius arm housing.

The cone bump stop, shims, spring and support plate assembly are next. After waxoyling everything, including up inside the chassis, fit and lock up the studdings as before and wind up the spring, fitting the six bolts before removing the studdings and replacing with the final four bolts. Last but not least, jack up suspension to remove packing and fit top link bump stop. Repeat for the opposite side of the vehicle.

I hope this has been informative and interesting and should anyone else have some additional ideas or improvements, let Overdrive know so as we can all benefit.

Best of luck

Mike Coleman (Mr. Modification) ***March 1998***

SAFETY BELTS

I share Steve Shone's concern for safety as detailed in his article of January 1998. We quickly fitted front seat belts after buying our '105S' eleven years ago and have fitted a variety of rear seat belts with the birth and growing up of our three daughters.

The youngest has always sat in the middle of the rear seat where the child's seat fits conveniently high on the centre arm rest. This has been a popular position as it gives a good view of the road between the occupants of the front seats (i.e. mum and dad!). In this position the two lower belts pass between the seat cushion and squab and are anchored near the transmission tunnel. The two upper straps have been secured by drilling through the parcel shelf into the two brackets beneath (I would welcome advice as to whether these are really up to the job). As the children have grown older I have had to devise increasingly complicated arrangements of angle iron to spread the load and provide fittings for three separate belts. I was never completely satisfied with these arrangements and wondered how well they would have coped in a serious accident.

I was pleased to discover when attending the Guild spares day in 1995 (I think) that Phil Beton had three sets of original Rover Co. rear seat belts which I snapped up right away. I bet not many members know that Rover listed seat belts when the P4 was in production. They are listed as Part No. 316921 made by Irvin. I was very fortunate to obtain these as Phil subsequently had numerous enquiries in response to an advert which he had earlier placed in Overdrive. Whilst it is marvellous to have the original belts, even more useful are the fittings and fitting instructions which came with them. The two lower anchor points are in the inner wing and near the transmission tunnel in roughly the same positions as described by Steve Shone in his letter. These are located with 'U' bolts and re-inforced steel plates.

A purpose made bracket is supplied for the upper anchor point (for the shoulder strap). These are 'handed' so each kit has two, thereby allowing fitting of the belt on either side of the back seat. These brackets attach to the boot hinge support using the existing bolts, thus no drilling (or angle iron!) is needed at this point. A small modification is made by cutting a slot at the front of the parcel shelf allowing the shoulder strap to pass neatly through and over the back of the seat. This bracket seems to me to be the key to safe and neat fitting of rear seat belts and should provide an anchor point for any three point static belt.

I have had a word with Peter Rabjohn who is willing to look into reproduction of these brackets by the Guild if there is sufficient demand. We have agreed that members should in the first instance contact me if they are interested.

I am able to provide photocopies of the fitting instructions (Part No. 316922) for rear seat belts and likewise (Part No. 316935) for front seat belts. Incidentally Rover also supplied a special plate to provide anchorage for the shoulder straps of the front seat belt on the B/C post. This neatly fits behind the trim panel.

If you would like the fitting instructions for the front or rear belts please send a large stamped addressed envelope (each set is four pages) plus one first class stamp (two if you want front and rear) to cover the cost of photocopying, to myself at 10 Elfort Road, London, N5 1AZ. Do let me know if you are interested in the brackets for the rear belts so that if there is sufficient demand I can ask Peter to look into reproducing them for the club.

As well as contributing to the safety of the rear seat passengers in the event of an accident, seat belts also locate the children in one place and give them the message that travelling by car (whilst being great fun and particularly rewarding in a P4) is also a serious business. Having developed a great interest in the literature of the 50's and 60's when I was editing the Year Book, I was interested to see that one of the strongest arguments against the use of seat belts was that children would not be able to keep still and would not put up with being confined. We have not found this a problem with our daughters, perhaps because we started when they were too young to know any difference.

Best wishes and safe driving.

Bob McGavin ***March 1998***

IMPROVING AUNTIE'S REAR END

Here is my modification to the boot hinge mechanism which instead of relying on the inner wheel arch as a mounting, now uses the boot floor as it's anchor point. Some years ago I replaced the rear inner arches of my P4. I couldn't bear the thought of introducing seven bolt holes into perfectly watertight arches and so was born the idea for this modification.

Have two L-shaped brackets made up out of ⅛" mild steel plate – 2" base, 5" in height and 12" in length. Now trim back one end of each bracket to allow for the rearward rake of the rear seat bulkhead. With the bracket pushed snugly up against the wheel arch, drill three holes through the boot floor at points 1", 5" and 8" from the rearmost end of the brackets. It will be noted at this stage that the brackets will cross over a boot rib pressing. Two spacers of ¼" mild steel needs to be placed either side of the rib, under the bracket. Under the boot floor will be needed a plate of mild steel with three nuts welded to its underside lining up with the three bolts coming down from the boot.

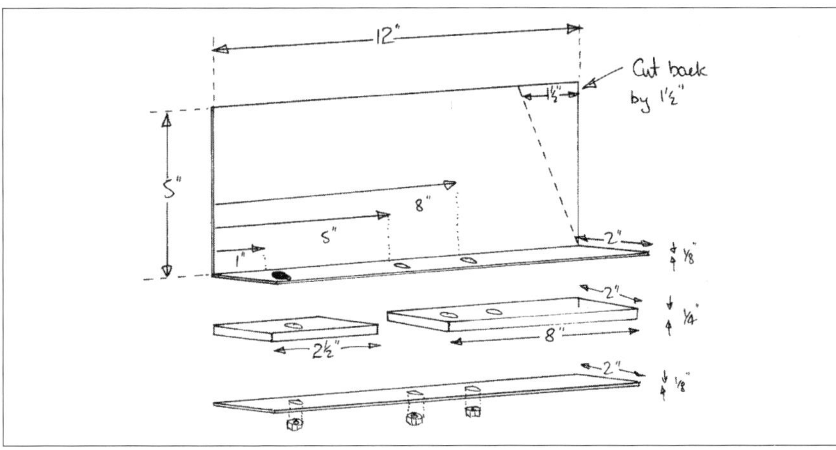

Now before marking out the remaining bolt holes please take out the hinge and trunnion bar and make sure that the pin on one and the hole in the other are serviceable. the pins on both of my hinges were almost "sawn" through! So having rectified any problems here, relocate the hinge and then mark out the four holes that will secure the vertical "leg" to the new bracket. It will now be necessary to take out the new bracket in order to correctly position the trunnion bracket.

Position the trunnion bracket so that it is resting on the foot of the new bracket. Mark and drill out the lowermost hole at a distance of 7⅛" from the nearest of the first group of holes previously drilled. Now tilt the trunnion bracket to an angle of 77° from the horizontal, mark and drill out the remaining two upper holes.

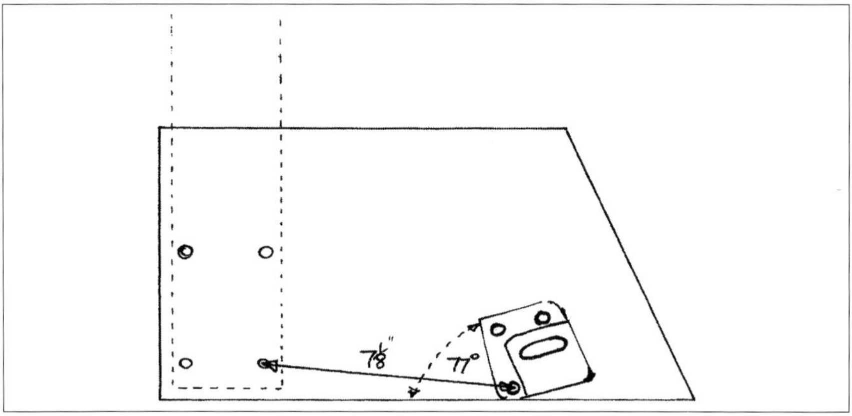

Bolt the assembly together applying plenty of grease to all moving parts, re-tension the compression springs and there you have it. The boot lid will now leap open to its maximum height and stay there. the stresses have been taken out of and away from the wheel arches and the whole thing will remain completely concealed by the original boot trim. Yes it will all fit snugly behind. Throw away that broom handle!

Any member wishing to evaluate the modification please give me a ring 0151 334 3858 and you will be more than welcome to come around and examine first hand.

Keith Hunter ***July 1998***

CORROSION IN BRAKE BLEEDERS

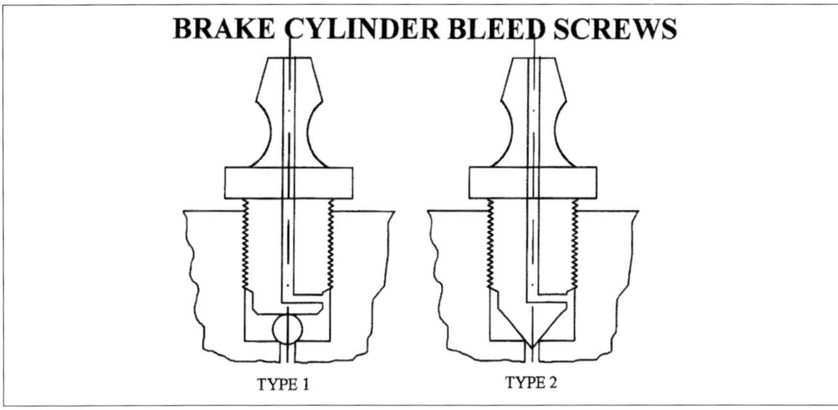

When I last carried out some work on the P4, I had broken down the brake system and it was necessary to bleed the hydraulics. I pressurised the reservoir to blow my system through and the air came out easily enough although it is a job I hate; the fluid is really the most unpleasant of substances.

I sustained pressure on the pedal for a while to check that there were no leaks and then went for a road test. After a few miles the brake fluid level light came on; my first thought was that the micro switch on the hand brake was set a little fine, but on checking, the lever did not put it out, and so I stopped to inspect. I had lost most of the container of fluid which was now around the back plate of the rear offside wheel.

The problem eventually turned out to be corrosion of the little ball bearing under the bleeder screw, which having been disturbed when the brakes were bled, had seated across the corrosion and was not sealing. I replaced the balls, and all is well. I hate to think where I would have ended up if I did not have the system with a fluid level warning.

There are two types of bleeder fitted to P4's and these are illustrated in the drawings here. It is possible to fit the conical screws in place of the ball and plain screw type, but the ball must be removed - your magnetised screwdriver will get it out. I stayed with the same arrangement as original, but put new balls in; either is technically satisfactory. I suspect my failure to detect the leak on static test was because the servo was not on and probably

I did not get the pressure high enough to reveal the problem.

As an aside, another P4 owner told me that he was fed up with the bright warning light burning his eyeballs whenever he put the hand brake on and he had disconnected the warning switch. I had to point out to him that the main reason for the hand brake warning is a routine bulb test function without which you could lose fluid and not get a warning if the bulb has failed. The servicing instructions also call for a routine check on the float switch.

Barry Kensett ***July 1998***

REAR AXLE PINION OIL SEAL REPLACEMENT

I would say that at least 80% of the P4's I have been privileged to work on, have a most common complaint – oil seepage from the rear axle pinion seal. Most of them appear to have been leaking for many years, and quite often, the axle casing is at least an inch bigger than it should be! One good thing about this of course, is that it stops the casing from rusting up and at the same time keeps the rear section of the chassis in a 'gooey' but protected state. The bad side of course, is the fact that you cannot tell how much oil is left inside to lubricate those most stressed and vital parts, unless you make very regular checks.

To replace the pinion seal, which of course, is the only answer, is not a difficult job, just messy and slightly awkward if you haven't got a hoist or a pit.

The first job is to drain the oil from the unit, this could range in volume from an eggcupful to 3½ pints. Then to make life as easy as possible, get the car jacked up prior to work. It helps to elevate the front of the car first, either on sturdy axle stands sitting under the front of the radius arms, or ideally, with railway sleeper-type blocks under the front wheels, approx 6" between tyre and ground. Whenever possible, leave the road wheels on the car, and in this particular job, this can be done, because they will not get in the way. You can then jack the rear of the car up, and support it, using axle stands under the axle casing, next to the road spring supports.

Once the car is jacked up fairly high off the ground, and you are happy that it is well supported, you can then clamber underneath and start by cleaning the area around the pinion seal housing. When this is spotlessly clean, you can then start dismantling. It is most important that everything is clean before you remove the seal housing from the front of the differential, because debris could drop into the unit and thus quickly destroy the bevel pinion front bearing.

Undo the rear propshaft flange from the driving flange on the differential. Then remove the split pin just inside the driving flange and undo the nut securing the flange, handbrake fully on! This nut should be torqued to 85lb/ft but sometimes can take some shifting.

Slide off the flange and then knock back the tab locks from the six bolts around the housing, and then undo the bolts and remove the housing. Whilst you are fitting the new seal, it is wise to wrap a piece of clean rag around the diff so that dirt is kept away from the exposed bearing. Clean the housing, then tap out the old seal. Meticulously clean the recess in the housing, smear hylomar around the OD of the new seal and with the use of a vice and two flat pieces of wood, gently squeeze the seal fully into the housing with the open end facing inwards.

Making sure the mating faces are clean, smear hylomar over the face of the pinion seal housing. With a smear of vaseline on the flange, push it on to the splines of the bevel pinion, replace the washer and the nut and gently nip up the nut, this should hold the housing in its central position. Then replace the lock tabs and bolts and tighten the housing to the axle body. Make sure the prop shaft bolts are fitted in the flange and tighten the pinion nut to approx 85lb/ft aligning with a relative hole for the split pin. The propshaft can then be replaced and bolted to the flange.

The only possible snag in doing this job, is the fact that the flange diameter around which the seal runs, may on occasion, be 'grooved' with wear, right where the seal runs. Even worse, it can be pitted in this vital sealing area, and quite honestly is only fit for the knacker's yard! Not all is lost though, because if you are lucky, you may be able to get hold of a double lipped seal, which, as its name implies, seals not only in two places, but either side of the original 'track' of the single lip type. If you cannot get hold of a double lipped seal, you will have to find a donor flange, with a good surface finish for the single lipped seal.

Stan Johnstone ***July 1999***

P4 HANDBRAKE MECHANISM MOD

You may be interested in the attached sketch which shows a very simple modification to the P4 handbrake operating ratchet mechanism, to overcome the well known problem of the pawl tending to pull sideways as the brake is applied, and causing the pawl to sometimes "jump back" a notch, resulting in a poor handbrake hold.

The mod is not intended to mask excessive wear in the various bearings or pawl, which of course must be maintained in good order; but having applied this mod; (which consists of a steel plate of 18-20 SWG steel, with holes drilled to fit over the studs which take the usual fixing washers), I have experienced no further problems over the past year. In fact it is now a pleasure to be able to park on steep hills again with complete confidence! (The steel plate may be adjusted with packing washers if necessary such that a nice 'sliding fit', without any undue sideplay is achieved and should, of course, be maintained well greased).

C.H. Ford ***1990 Yearbook***

P4 HANDBRAKE MODIFICATION

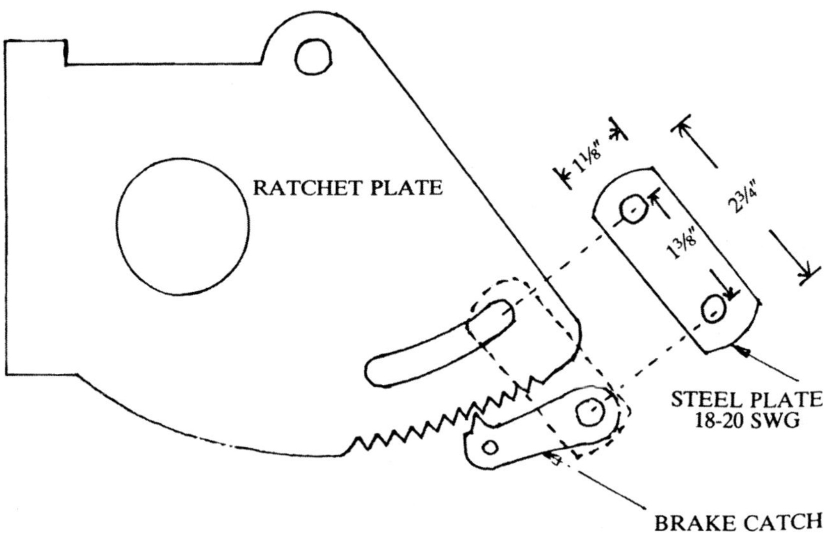

PLAN YOUR MAINTENANCE PROGRAMME

Your trusty P4 will remain so, provided you treat it with respect and carry out the necessary maintenance. It helps if you can stick to a programme so that you don't end up having to spend a week just catching up on the various jobs that need to be done. This is to say everything does not need to be done at one time. You can spread out the work to be done according to your mileage, often I think, unfortunately, governed by the insurance companies.

By way of a suggestion I would recommend the following procedure, bearing in mind your particular mileage. If for example you cover 5,000 miles per annum you will need to do two engine oil changes a year. Whereas if you only do 1,500 miles a year one oil change would suffice. That is to say, the recommended period being 3,000 miles the oil would remain in the engine for two years. This is not good because oil deteriorates even just standing, so it would therefore be safer to change the oil once a year on a 1,500 miles a year schedule.

Always change the oil when the engine is at full running temperature i.e. after approximately a 15 mile run. Obviously the filter will need to be replaced with a new canister at every oil change.

Clean the spark plugs on a yearly basis regardless of mileage (unless you are doing in excess of 10,000) and reset the gap. Change the contact breakers (points) after about 5,000 miles and if they look particularly sooty replace the condenser as well. Reset the points gap and check the ignition timing. Examine the plug leads at each end to see that no corrosion has taken place and check that the contact is clean between the suppressors and spark plug dust cover (not fitted on '95' and '110').

Tappets should be checked and readjusted if necessary at 5,000 mile intervals. These are best done with the inlets hot (about 15 mile run) and the exhaust 'overnight' cold.

The coolant should be drained and flushed every two years regardless of mileage and refilled with a strong solution of antifreeze (about 7 pints of antifreeze). At this point the thermostat (must be bellows/alcohol type) should be removed and checked in near boiling water to see that it is opening fully.

The swivel pins, steering box, and steering idler levels should be checked every 5,000. The swivel pin reservoirs are best checked using a dip stick made from ⅛" dia wire (an old wire coat hanger is ideal) and a level should be marked on the rod at 6¼". If there is a considerable loss of oil in the pins this will need to be replenished on a more regular basis, or new seals fitted.

Check front brake pads (late P4's) before every M.O.T. and replace if worn down to about ⅛". Also prior to M.O.T. remove rear brake drums and examine slave cylinders for fluid seepage, if any. Remove plate from rear of cylinders and clean the sliders and rollers on the handbrake mechanism. Regrease. Spray all pivot points of hand brake with WD40. Adjust brake shoes.

Every two years change oil in back axle and gearbox. This level should be checked on the dipstick every couple of months.

Check brake fluid level every week when car is used regularly. Once a year spray door locks with WD40.

It helps to keep a record of maintenance and mileages and with a bit of careful planning all these jobs can be spread out through the year.

Stan Johnstone ***1991 Yearbook***

BRAKING POINT

As Technical Advisor for the Guild I get asked many questions on the mechanics of the P4. By far the most popular query is concerning bleeding the brake system. Many members have found that it is not always a simple case of pumping the air out into a bottle. The most common problem occurs after refitting the master cylinder albeit a replacement unit or the original with new seals. What many fail to do is to prime the master cylinder with fluid before fitting. This is done by carefully pouring a small amount of brake fluid into the inlet hole on the top of the cylinder and gently actuating the push rod until fluid is flowing from the outlet hole. Be very careful of your eyes when carrying this out!

Once the cylinder is full of fluid refit to the car. Always use new fluid on refilling the system and bleed the brakes in the normal sequence i.e. back nearside, back offside, front nearside and finally front offside. Of course the most easy and effective way to bleed the system is to use an 'Easi-bleed' kit made by Gunsons. With this you can do the job on your own (without bothering the wife) and not touch the brake pedal at all. The kit simply takes pressure from the spare tyre (let down to about 20 p.s.i.) and pressurises the fluid reservoir. With the kit you get a selection of caps to fit the reservoir. There is one very near to the Rover size but I have used a spare one from a

P4 and drilled a hole in it to fit the adaptor supplied. This is safer because you do not want to run the risk of spraying brake fluid all over the off-side wing! The 'Easi-bleed' costs about £12 from Halfords and is well worth the cost.

Another question often asked is how to get the handbrake to D.O.T. standards. Generally this is achieved by checking that the rear slave cylinders are free to move on the back-plate. Invariably they are seized solid to the plate and will not allow the brake shoes to centralise. This is easily rectified by removing the rear brake cylinders and wire brushing the brake back-plate inside and out along with the small plate which the nuts and special spring washers tighten onto. Grease both sides of the back-plate and fit the cylinders systematically, tightening the three nuts until it is held against the back-plate but is still able to slide up and down in the slots. Clean and replace the hand brake drawlinks, tappets and rollers and grease with copper-slip. Check that the balance lever on the rear-axle is set to the correct angle and not pulling against itself and you should then have a good handbrake to keep you and the tester happy.

Stan Johnstone ***1991 Yearbook***

REAR ENGINE MOUNTING – REPLACEMENT PROCEDURE

Of all the technical enquiries I receive, one of the most often asked is how to replace the rear engine mounting. Yes, it is called the rear engine mounting, not the rear gearbox mounting as it is often erroneously called, because, whilst it is the only support the gearbox has, apart from being bolted to the back of the engine, without it the engine would have only the front two mounts to hold it in place.

This very important component works very hard for a living and has an immense stress imposed upon it. Its greatest enemy is oil which destroys the molecular structure of the rubber encasing it. Rear crankshaft oil seals and top rocker cover gaskets are the main culprits. When it is of an age, the mounting loses its strength, allowing the engine and gearbox to become liberated developing that well known shudder, particularly when pulling away from rest or reversing.

Make sure that the front engine mounts are in good order first. To check these, gently try to jack up the engine from the front whilst watching the mounting rubbers. Once under load, they should show no sign of being split or broken. If you have established that they are alright, you will have to replace the rear engine mounting. Tackle the job thus:

i) Remove the front seat(s), carpet, gear-change mechanism and gearbox cover. You should now be looking at the gearbox in all its glory.

ii) Now get under the bonnet, remove the top hose, throttle linkage from the carburettor, and separate the exhaust manifold from the down pipe.

iii) Ensure the car is safely supported for you to clamber underneath, I use heavy blocks of wood approximately 12" high under the front wheels as I always like to have the car standing on its own feet as it were. You can, of course, use axle stands. Make sure that the rear wheels are chocked and please, please, take the utmost care, many people have been squashed, many fatally, whilst working on the underside of cars.

iv) Drain the gearbox oil . . . . . completely.

v) Under the car, remove prop-shaft from rear flange of the gearbox/free-wheel (dependent on model) and remove rear engine mounting bolt from mounting and chassis. Undo the clutch cross shaft from the side of the gearbox and (if fitted) remove the overdrive solenoid from the overdrive unit.

vi) Gently jack up the engine at the point where the gearbox bolts to the engine, I use a block of wood with a vee sawn into it so that it stays central during lifting, until the top studs holding the gearbox to the rear of the engine are just touching the apex of the aperture in the bulkhead.

At this point the gearbox will be sitting up into the interior of the car.

vii) Undo the nuts and bolts holding the extension casing to the rear face of the gearbox. Ensure that you only split it at this point and not where the overdrive or free-wheel joins the extension casing. (It is the extension casing that houses the rear mounting on overdrive models, the tail-shaft housing on non-overdrive models and in the free-wheel unit on models so fitted).

On free-wheel models the ferrule holding the free-wheel cable must be removed to allow the unit to be pulled away from the gearbox.

On overdrive models it is vital that the isolator switch is withdrawn from the extension casing. Make sure that the plunger inside it is removed, this is activated by the 3rd and 4th speed selector shaft and if it is not removed you will find it impossible to slide the unit back on to the gearbox when you have fitted the mounting and are putting it back together.

viii) Slide the unit away from the gearbox.

(As an aside, I always remove this unit first when removing a gearbox as the two lumps are, obviously, lighter separately. It makes the refitting of the gearbox easier. In my younger days, sigh, I used to do it in one lump. Now I like to make life as stress free as possible).

ix) Spread newspapers on the dining room table before you take the unit inside to remove the old mounting. Cut out the rubber, using a very sharp knife and hacksaw, a slot along the width of the outer sleeve until the blade just meets with the aluminium of the casing, the sleeve will then literally lose its grip and will be easy to tap out. You will definitely damage the casing if you do not split the rubber first.

The mounting is an unbelievable seven thou interference fit which is tighter than anything I am allowed to mention in this publication. The outside diameter of the mounting is 2.255" plus .002" and the hole in the casing is 2.250" minus .002", so to fit these two into each other would be impossible at a common temperature.

x) You should place the new mounting in a plastic bag and place it in the coldest compartment of your freezer for, preferably, 24 hours prior to fitting. Clean out the bore of the casing and gently heat the area around the hole using a blow lamp but make sure you keep the heat localised. When the aluminium is really hot insert the frozen mounting with the eccentric hole on the centre line and to the base of the unit. Quickly tap in the mounting using a copper mallet until it is fully home. (Strong industrial gloves are needed to protect your hands from the two extremes of temperature).

xii) Allow everything to cool down, including yourself, and you are then ready to put the whole lot back together again. As they say in all the best workshop manuals, "sling it all back together in reverse order".

A lot of work for such a small component but well worth the effort when, once again, you pull away from rest as every Rover should.

Stan Johnstone ***1997 Yearbook***

BARREL LOCKS BY THE TECHNOCRAT

Good old Wilmot Breeden, some of those locks on our favourite motors have been working for a very long time. But, have you ever had to replace a barrel and are frustrated that you now have to carry one key for the door and one key for the ignition? Would you like to have them both the same with no greater cost than a little patience? Then read on for the technocrat has the answer.

How does it work?

In an efficient lock in good repair, the barrel proper is able to rotate either a quarter or a half turn within the outer casing but is prevented by doing so by the five tumblers that are spring loaded to engage in a slot within the outer casing unless the correct key is inserted. Each tumbler has a rectangular slot and one end, the "acting face", is at a calibrated distance from the tumbler end.

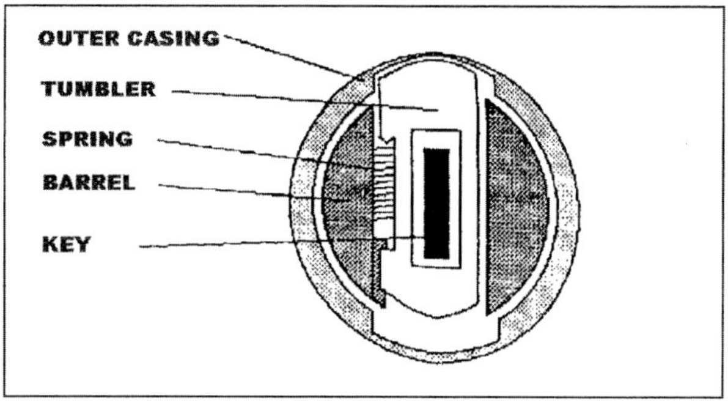

Figure 1

When the correct key is inserted, the part of the key in line with each tumbler exerts pressure so that the tumbler is flush with the outside of the barrel on each side so that the barrel is able to rotate. See Fig. 1

An incorrect key will move the tumbler but either too far or not far enough thereby preventing rotation.

So, what goes wrong?

The most common problem is with sticking. Oh, a squirt of WD40 will cure it for a little while but it will wash out the actual lubricant and it will freeze in winter. The body of the barrel is made of our old friend Mazac (zinc alloy) and the tumblers are of brass or plated steel. Whaddya get with mis-matched metals? Electrolytic corrosion. So, you need a coating of grease to keep them apart.

Removal for lubrication

The barrel is held into its housing by a latch which resembles an extra tumbler.

Once the lock has been removed a small hole will be visible into which can be inserted a scriber to push back the latch. Occasionally, a barrel will fall out because the latch has fallen out or is simply stuck.

On dis-assembly, carefully wash all the parts, grease well and re-assemble.

Making a barrel fit a different key

Right, you have obtained a replacement barrel and you want to make it fit your old key. It will help if you have a supply of old lock barrels, or you may get away with fitting the tumblers from the barrel you want to replace.

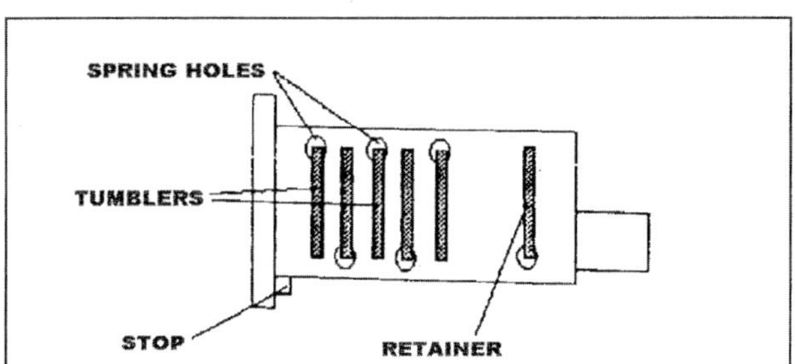

Figure 2

Wash all the bits and remove the tumblers.

Locate the side of the barrel where the spring holes are visible, we will call this the top for the sake of convenience. Fig. 2

The tumblers can now be pushed out from the bottom with a small screwdriver. Work over some sort of container, you are unlikely to find a spring if it falls on your garage floor. Repeat the process on the old lock or any scrap barrels you may have to supply a set of tumblers.

Now, insert the key and one tumbler at a time by inserting them upside down in the bottom of the barrel. You are looking for those with action faces that fall flush with the barrel. Fig. 3

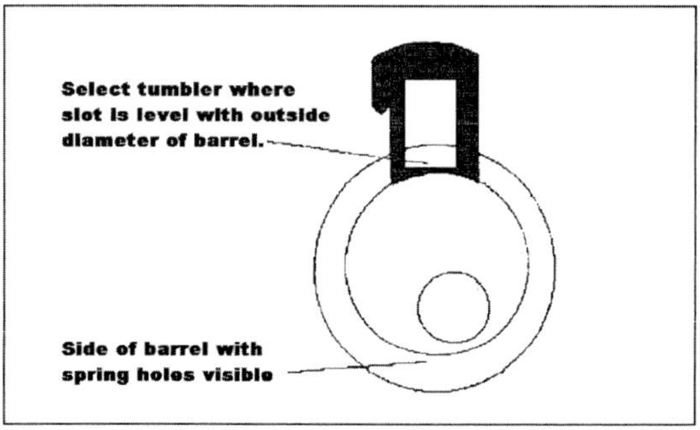

Figure 3

Each time you find a tumbler, remove the key and fit the tumbler to the barrel. A tip is to smear the tumbler with grease and insert so that the spring slot is half in. Then insert the spring, lower end first, gently compress it and engage the top on the triangular projection in the slot. Push the tumbler in and when the key is once again inserted the tumbler should, of course, be flush. Repeat this until all the tumblers have been inserted.

Smear the assembly with grease and re-fit.

It really isn't as complicated as it sounds and is much easier to do than to write about! It is possible to leave out a tumbler but this will be at the expense of security.

Anon. ***Yearbook 1997***

IS AN UNLEADED CONVERSION NECESSARY FOR A P4?

Firstly, only you can decide whether to do any remedial work, or just wait until the issue is forced upon you, and you just have no choice but to address the problem. Personally, I think the latter option is the best bet. As I have said before, why fix it 'til it's broke? I, like many P4 drivers, have taken the plunge and have been using unleaded fuel without preparing the engine in any way. If you are going down this road, I think the prudent thing to do though is to make regular checks on the tappet clearances, especially the exhaust valves. These should be set very carefully before you start using unleaded. These exhaust clearances are best checked when the engine is 'overnight' cold, but the inlets must be set or checked with the engine at full working temperature, i.e. after at least a thirty mile run. By monitoring these clearances at regular intervals, you can keep an eye on any valve/valve seat recession that may or may not be occurring.

TOOL FOR DRAWING IN EXHAUST VALVE SEAT INSERT

The seat insert has an interference fit in the cylinder block of .005 in. to .007 in. (0,13 to 0,18 mm.) and must be carefully pulled into position to prevent shattering.

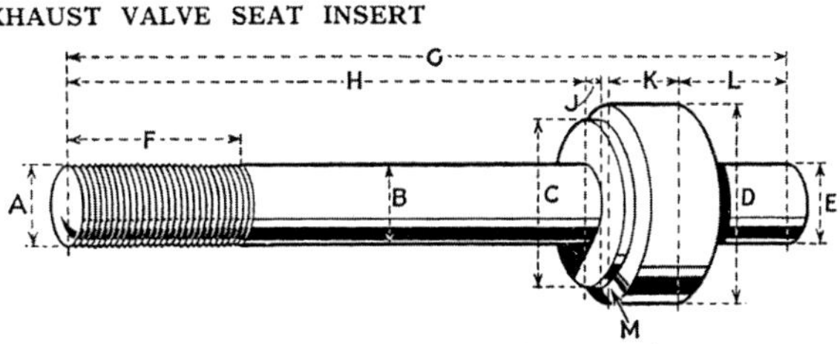

Fig. Y–5. Tool for drawing in exhaust valve seat insert.

A—$\frac{9}{16}$ in. B.S.F.	E—$\frac{9}{16}$ in. (14,3 mm.)	J—$\frac{1}{8}$ in. (3,2 mm.)
B—$\frac{9}{16}$ in. (14,3 mm.)	F—$1\frac{1}{4}$ in. (31,75 mm.)	K—$\frac{1}{2}$ in. (12,7 mm.)
C—1.137 in. (28,87 mm.)	G—$5\frac{3}{16}$ in. (131,8 mm.)	L—$\frac{3}{4}$ in. (19 mm.)
D—1.370 in. (34,8 mm.)	H—$3\frac{3}{4}$ in. (95,2 mm.)	M—30° to suit seat.

If, however, you would prefer peace of mind and feel inclined to embark on an unleaded conversion, I will relate to you the correct way to go about it. The main components that will have to be changed are the exhaust valves themselves. The main Rover partsmen now have these in stock (but for how long I don't know, so it might be worth putting a set on the wine rack!) they have a high stainless steel content and are up for the job. The valve seats for the exhaust are capable of coping without lead, but must be in first class condition, i.e. no surface cracks. If they show any signs of wear they will have to be changed, and if so, you will invariably have to change the exhaust guides too, because once the new seat is installed, its face will have to be cut and the cutting tool uses the guide as its pilot. An old worn guide will not allow a concentric face to be cut. Removing the guides is relatively easy, using the appropriate drift (see diagram), but seat removal is not so easy. Ideally the seat will have to be weakened by using an air drill with a small grinding burr fitted. Extra care must be taken in this procedure, working at the top of the seat furthest away from the cylinder bore. Once the seat has been virtually ground through, it can then be broken using a good quality, sharp chisel. Wear your goggles for this job! The seat recess can then be cleaned out thoroughly, ready for the new seat to be fitted. These seats are very hard and of a brittle nature, so the only way to fit them is using a pulling tool as shown in the diagram. The new seats should be pulled in when the guides are removed, and once fitted, the new guides can be drifted in ready for the new seats to be cut. You can then lap the new SS exhaust valves into their mating seats.

You can then concentrate on the other half, namely the cylinder head. I have covered this topic in Yearbook 1992, so will not go over it all again. If you haven't got that particular article, send me an SAE and I will send you a photocopy of it.

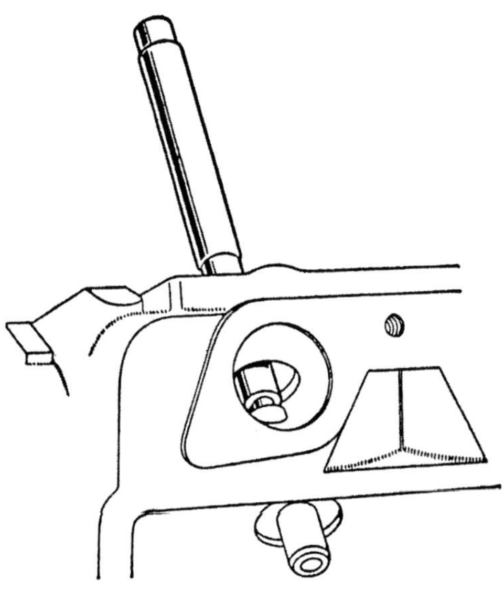

Removing exhaust valve guide with drift

To sum up, as regards the head, it is essential to have the face surface ground to zero flatness and most certainly not 'skimmed' as most companies would do, as it will not end up flat and the surface finish will not be fine enough to prevent 'gassing'.

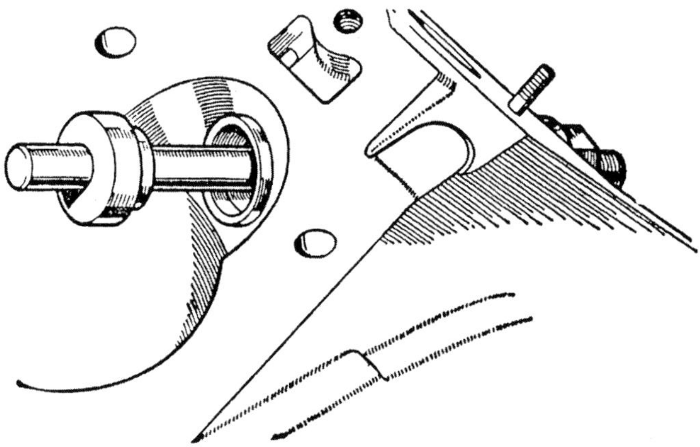

Fitting new exhaust valve seat

It will be well worth replacing the inlet valves guides and recutting the seats. Inlet valves are not expensive to buy as opposed to exhaust valves, so it makes sense to buy new and lap them in to the recut seats. Do not forget to fit new 'O' seals to the inlet guides. When replacing the inlet rocker gear into the cylinder head, it is wise to replace the inlet rocker spacers (x3), so that the rockers sit centrally over the valve stems.

Once all this preparation has been done, you are then set for some unleaded motoring, but before you do, you will have to give some thought to the cooling system.

The engine will run hotter on unleaded fuel, but this shouldn't be a problem if the system is up to par. The P4 engine in fact, was overcooled if anything, but that could be as long as fifty years ago, or as short as thirty-five years ago, depending on the model, so it is fair to say that some silting, and thus congestion, will have occurred. for complete peace of mind, it would be an opportune time to remove the core plugs along the side of the engine block and with the use of a power hose, thoroughly flush the block, paying particular attention to the rear of the block. Since the P4 engine sits lower at the back than the front, it is very prone to silting, and being furthest from the fan does tend to run much hotter at the rear, because of the lack of coolant circulation going round the rear cylinder jackets. A good poke round with some sturdy wire should encourage a bit of descaling to take place, with the use of the power hose. When you are happy that the block is a lot cleaner than it was, new core plugs can be fitted.

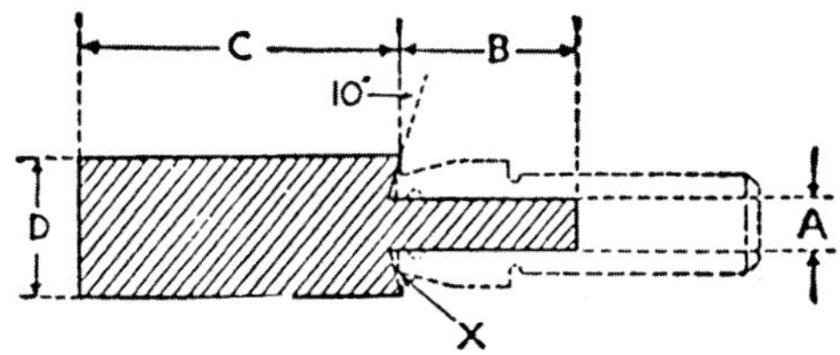

Fitting tool for exhaust valve guides.

Dimensions and material. A=.340 in
Mild steel. B=1 in. C=$1\frac{3}{4}$ in. D=$\frac{3}{4}$ in. X=Point of impact.

Radiators are invariably furred up and therefore not as efficient as they should be, so a radiator specialist can run a flow test on it and at the worst, may advise a new core. Expensive maybe, but if it means you can carry on driving your P4 for many years to come, it's money well spent.

That's it, that's all you need to do to your Rover, in order to drive it way beyond the millennium.

Stan Johnstone ***1999 Yearbook***

INSIDE INFORMATION

I have never been particularly impressed by the handbrake on my P5 which although similar to the P4 is operated by an umbrella handle and cable rather than a proper lever. I assumed the reason for this poor performance was the bedding in of new linings and the lack of a proper lever.

Recently however, things took a turn for the worse after a rear brake service when the hand brake failed to hold the car on hills – necessitating working three pedals with two feet to avoid rolling back into the car behind.

The shoes had been aligned in the drums and steady posts adjusted in accordance with the workshop manual and all the operating rods and levers were correctly adjusted. Fortunately Colin and Anne Blowers visited one afternoon so I took the opportunity to get a fresh mind on the problem. After much investigation Colin noticed that the nearside handbrake draw link was pulling through to its stop when the handbrake was applied although the drum could still be rotated.

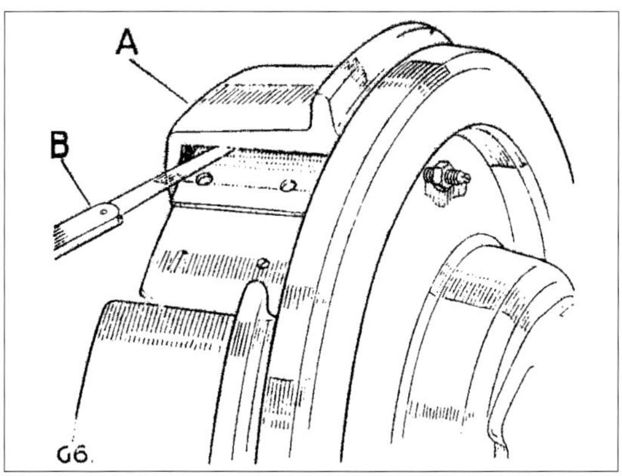

Again the drums were removed and Colin realised that the shoes were lying close to the backplate and as a result of this the linings were only contacting the drum on this outside edge. A readjustment of the steady posts resulted in a vastly improved handbrake and incidentally much reduced footbrake travel.

Having originally adjusted the steady posts in accordance with the manual I cannot understand why the shoes were so far out of alignment. I have had no problems of this nature with my P4.

I have since cut a section out of the circumference of an old drum which when fitted in place of the normal drum enables a feeler gauge to be used on the inner and outer edges of the lining thus enabling the steady posts and shoes to be adjusted with great accuracy. I now have a handbrake almost as effective as the P4.

Derek Humphreys ***2000 Yearbook***

BACKLIGHT

This photo shows how the glass frame seal, frame and body seal fit together. The whole assembly complete with frame and body seal is pulled against the edge of the body recess by clips and shims screwed to the car body from inside as shown later.

Whenever I have sought advice about removing the backlight, the almost unanimous reply has been “Don’t mess with it”. Unfortunately for me this advice had come too late. So for those of you who must, here are some tips.

There are definitely "two levels of hell" when it comes to the backlight.

The first level of backlight hell is when you remove the frame complete from the car, fit a new body rubber to it, and then refit the whole unit back to the car. When removing the backlight from the car always make a note of how many shims are used under each clip, since you are likely to use the same number when re-fitting it.

The worst problem that I found during re-fitting was that the new seal was a different section to the original. At first I thought that the differences were not significant, however the combined effect of the shallower curve at the top left where the frame presses it against the car, meant that the feathered edge would not sit against the body no matter how I tried. At the National I did some research, and ascertained that all the sections on offer by the spares men were essentially the same. I also talked to someone who had used one of the rubbers who said that he'd had to resort to gluing the edge down.

When fitting this seal to the frame it is easiest to hook the heel into the square part of the frame slot all the way round the frame first. Lubricate the crevice between the back of the heel and the frame (I used special Hellerine rubber lubricant for this, which is available from Electrospeed. You will have to phone them, because their website is awkward to use! The stock number is 254-62063D). Carefully work the seal into the frame with a blunt screwdriver.

This picture isn't very clear, sorry.

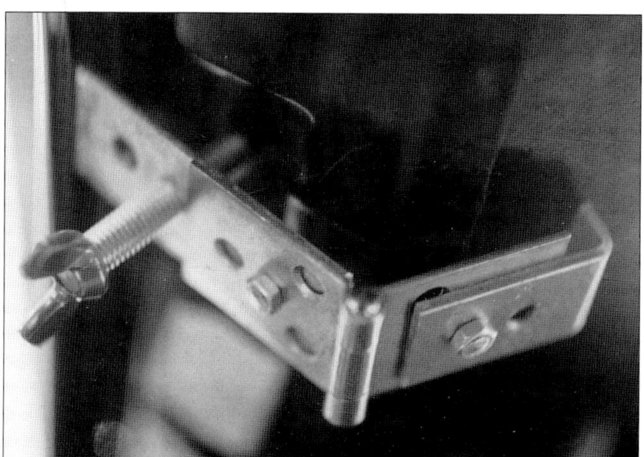

Here it is in position. . . . . .

. . . . . and here is one of the others in position. You can see one of the retainers that hooks under the edge of the frame, on the left of the clamp. Notice the shakeproof washer. This one doesn't have any shims behind it. The self tapping screws are of different lengths, so it's a good idea to poke something into the hole first to make sure there is enough depth available, otherwise you could distort the outer body skin as you tighten the screw.

The next problem I encountered was that the clamps have never been available so I decided it was probably easiest to make up a set. As you can see they're crudely made, but they work. I found that galvanised building ties are the cheapest source of suitable metal and the holes are conveniently just the right size to be tapped out to 8mm without extra drilling. The leverage is better when the wing-bolt is nearer the hooked end of the clamp, but the variation shown was acceptable. It is well worth making an extra couple of the top ones, because I ended up grouping the clamps together in places to get enough leverage.

The ones for the bottom at the sides are the same dimensions as those for the top, but need to be twisted. The bottom centre one is the most difficult to make, and has a prong that goes into the hole under the centre of the window aperture.

When fitting the backlight I put a bead of Evo-Stik Frame and Gutter Sealant round the lip of the body aperture before offering up the frame. This not only seals the join but allows the whole thing to slide into position more easily. It is a rubberised bitumen sealant and is closer to the original type than anything else I have found, also the excess can be removed with white spirit. I do not like silicone because if you ever have to take the window out again you can never get the stuff off satisfactorily without wrecking the paintwork. Getting the window to fit is a terrible job made worse by the fact that the retaining clips are screwed to the single inner skin of the body with self tappers, which may strip if excessive strain is placed on them. Use the clamps, levers, bad language and any other means at your disposal to pull the frame into position as described in the workshop manual. I would suggest getting the sides to fit nicely first, while watching the gap between the body and frame all the way round, and sharing out the free play, if any. You will find that if you pull the frame too far into the car the feathered edge of the seal will pivot away from the body. This may seem counter intuitive at first.

Once the sides are looking reasonable, put in some of the retaining clips in this area using shims as required tightening lightly, then turn your attention to the top and bottom edges. Do not be afraid to lever things carefully. Be prepared to start again. And again, until it looks right. When you have got it right, go out and celebrate!

The second level of backlight hell comes when you decide to dismantle the frame itself.

It is in two halves which are joined by straps in the glass recess with 4 screws each side. The bottom screws are likely to be rusted solid. The glass sits in a slot in the frame, cushioned, and hopefully (some hope) sealed by a U-section of pitch impregnated fibrous stuff. You can see this in the first photo at the beginning of this article. The sections of glass are joined by an H shaped rubber extrusion. Before you dismantle the frame, be warned, read the rest of this first!

For the U section the nearest seems to be section R250 from Woolies.

The inherent difficulty of assembling the frame being surmountable, the real obstacle lies with vertical strips. The original section is shown at the bottom. The section at the top is offered as a replacement by all the spares men. They claim that they've sold many and never had any complaints, but as you can see it is really intended for joining much thinner material, and I have never seen one used in anger. The really bad news is that I have not been able to find a source of anything that is better. It may be possible to fabricate something suitable out of T section, but in my opinion it is not currently possible to rebuild the backlight satisfactorily without re-using the original strips, so if you're determined to dismantle the frame, be very careful not to damage them, bearing in mind that they may have become brittle over the years.

Please e-mail me if you've tried the new section or found a better replacement.

The "Cheat Method" for Fixing Glass to Frame Leaks

With so many problems to overcome, I decided not to dismantle my backlight frame, but try and reseal it while still assembled...

I dug out some of the original fibrous seal between the outer surface of the glass and the frame, taking care not to scratch anything of course. Then I cleaned out the furrow with a brush. Next I ran two strips of masking tape round the window, one up to the lip of the frame and the other parallel on the glass, leaving a gap that I could work "U-Pol Tiger Seal" into using my finger without smearing it all over the frame or glass. This is a really first class sealer with superb adhesion, which hardens into a flexible rubber-like material. I carried the masking tape over the vertical strips so that sealant could be worked into the corners of the join to try and ensure a watertight seal. Removing the tape is a bit time consuming, but it really does save time in the long run. I found this method stopped all the leaks, but I think it is best used once the frame is back on the car and it has had time to settle, so that subsequent movement is kept to a minimum.

Roger Dealtry http://www.p4.org.uk **2000 Yearbook**

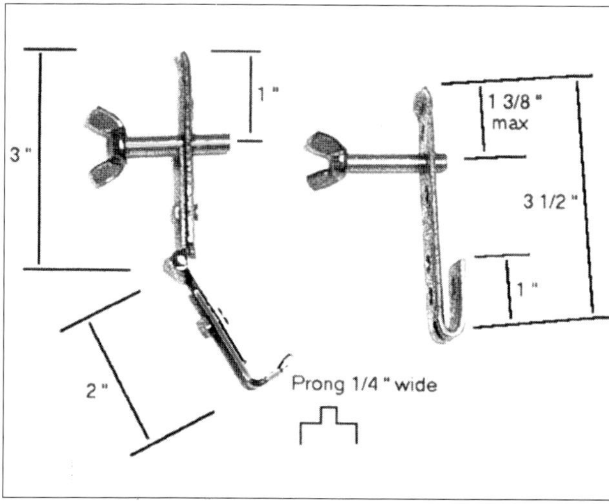

66. fittings tools for backlight

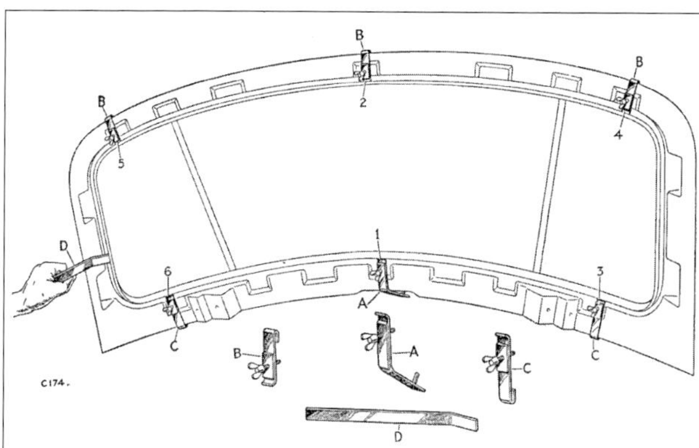

The complete set of tools as listed below are essential when fitting the three-piece backlight on the Rover car.

Part Numbers:	Models:
Type 'A' 314322 1 off	1955-57 '60', '75', '90' and '105'
Type 'B' 314029 3 off	
Type 'C' 314371 2 off	
Type 'D' 314385 1 off	

Numbers on illustration indicate the sequence in which to fit the clamp brackets when commencing to reassemble back light.

Taken from Rover Workshop Tools January 1957

I'VE BEEN FRAMED

Following on from Roger Dealtry's "Backlight" article in the Guild's 2000 Yearbook, I present you with a résumé of my experience of this dreaded task. For me the removal of my 1960 '100's rear window was not an option; the chrome frame was pitted, a bare metal respray was planned and the external rubber was brittle and breaking up. Having never dealt with the removal and replacement of a P4 rear window I employed a contractor to remove it for me. This was mistake number 1 as I did not see the number of shims required for each rear window bracket, more about this later. The next mistake was to employ the same contractor to refit the rear window, he had not done the job before and I ended up with the 3 glass sections sitting in silicon sealant and an incomplete job. I thought long and hard about what to do and fortunately chanced upon Roger's excellent P4 web site and the informative article on the rear window.

Refitting the glass into the Frame

Until I read Roger's article I had no awareness of the rubber u-section surrounding the edge of the three rear glass sections within the frame. On dismantling my rear window all I found was hard black stuff of an indeterminate nature and I assumed it was just old sealant. Anyway, Roger's article identified the R250 U-section from Woolies as being suitable so I went ahead and ordered the required length.

I decided that the dining table extended to its six foot length, suitably protected, was just right for the job and managed to convince my wife Jean of the merits of this approach; it was in the warm, the right size to work on and at the right height! Also closer to the kitchen for much needed refreshments.

The two new sections of the vertical rubber strip that separates the three glass sections had to be trimmed to the right length so I assembled the window frame to assess the required length. Unfortunately I measured the internal dimension and not the external dimension which is slightly longer and shaped to accommodate the profile of the chrome frame. Check a P4 before you attempt this to see what I mean. Because of this I have needed to prepare a small in-fill section for my car using some of the surplus I cut off. The R250 will not sit neatly around the corners of the glass and fit within the frame so Jean, using her sewing skills, cut small vee-shaped sections out of the rubber to help.

The logistics of aligning and sliding the 3 pieces of glass, 2 rubber strips and the R250 within the bottom section of the frame would have been a sight to behold. Both Jean and I struggled initially with the frame on the table and then on the floor to align and push the various components together, aided by Fairy Liquid for lubrication. I also put a small bead of the Evo-Stik Roof and Gutter Sealant into the frame as an added measure for water tightness. Unfortunately, this made matters slightly worse as the surplus sealant had to squeeze past the glass edged with R250 as the components were pushed together. Miraculously getting all the components lined up into the bottom section of the frame was not too difficult with the middle section of glass coming adrift just twice; this necessitated starting the process again.

Having dealt with the bottom half of the frame the time came for the top half. But at this point don't forget to fit the two metal strips that hold the two sections of the frame together. They fit internally within the frame with the four small screws being driven in from the outside of the frame. Fitting the top section of the frame was not too difficult as the glass sections were being held in more or less the correct position by the bottom half of the frame. However, on pushing the top section on the glass edged by R250 it became apparent that with normal force the two sections were about half a centimetre apart. This is where the project would have become a contender for the TV programme "You've Been Framed" as both Jean and me grappled with the framed window in all manner of positions on the dining room floor pushing with all our might. There is something about the shape of the rear window frame that makes it unsuitable for heaving it together. I tried to think of alternative means of forcing it together but in the absence of anything realistic continued to push down on the frame on the floor with all my strength, hoping that the glass would not crack.

When it reached the point where I could put the final 2 screws in to the 2 joining strips I was jubilant that this stage was complete, little did I know what was to follow. At this point all that was left to do before the next stage of the project (a nightmare come true) was to trim off the surplus R250 showing around the edge of the chrome frame with a Stanley knife (Johnstone gets in everywhere!).

Fitting the external rubber to the frame

On inspecting the new rubber it was apparent that it was too long and would need to be shortened and joined. So with the manufactured join at the centre-top we proceeded to fit the rubber to the frame with a plan to make a cut and join at the centre-bottom. Using Roger's advice I utilised a blunt screwdriver to ease the rubber into the frame with some Fairy Liquid for lubrication. This is easier said than done as whilst it is quite easy to insert the rubber along the straight sections of the frame it is very difficult around the curved sections, especially the tightly curved top corners. Again the dining table came into its own and with Jean's help holding the frame at the right angle the task was slowly completed. Does anybody know how Rover did this on the production line? There must have been a special tool. The join in the rubber was achieved using Superglue and after allowing it to set the last section was inserted into the frame.

Fitting the complete rear window into the car

Nothing could have prepared me for this stage in the project. Fortunately I had been in contact with Roger through his website link and not being particularly practically minded to make a set of clamps for myself I decided to ask Roger if I could borrow his. I will be forever grateful to Roger for the loan of his clamps.

Using the advice in Roger's article I put a toothpaste size bead of Evo-Stick Roof and Gutter sealant around the aperture and levered (using a variety of implements, mainly a large screwdriver), clamped and swore. Jean stood outside the car giving a view on how flush the window appeared with respect to the car's bodywork. This procedure was made more difficult because I had no record of the number of shims required for each bracket owing to the rear window being removed by the contractor. However, after much effort the window was in but it proved impossible to get the bottom near-side corner to fit neatly and flush with the shape of the surrounding bodywork. So I removed the whole frame! The amount of leverage necessary to pull the frame into the aperture also seemed excessive to me and I was really concerned that I was doing something very wrong. This being especially of concern when some of Roger's clamps started to bend with all the tension.

Having cleaned the sealant off with white spirit we had another go and this time the self tapping screw holes for a couple of the brackets stripped! So I went up a size of screw. Unfortunately, the amount of screwing and unscrewing stripped the holes again. Fortunately Rover provided enough space around the space where the brackets fit to enable a new hole to be drilled. Imagine drilling in this area with the potential to drill through the external skin of a car which has just had a bare metal respray! Unfortunately, for a second time it proved impossible to position the near-side bottom corner correctly.

I decided to undo the brackets, clean the frame up and seek some moral support and guidance from Roger. After a very long telephone conversation with Roger and further discussion with other Essex Branch members I decided that I would have another attempt. To my horror, on looking at the rubber seal, I noticed that cracks were beginning to appear at the corners! I assumed that this might have been as a result of cleaning up the frame with white spirit so I purchased a new seal and started the complete process again! Should white spirit have done this to the rubber?

This time I got as far as inserting the new rubber into the frame and a variety of events occurred that resulted in a few months of no work on the car. On hearing the disastrous tales at Essex Branch meetings one of the members, Jason Dorey, kindly offered to give me a hand. We agreed a convenient date, however, two days before this date I noticed that the corners of the new seal were starting to crack. So we delayed the task until I purchased another seal. I began to wonder if the rubber was not flexible enough to bend around the corners of the frame. During the intervening period I looked at Jason's P4, a 1962 '100' with its original window rubber. It was apparent that the original seals had a greater amount of external rubber to fit against the external bodywork of the car. I discussed the matter with a different supplier and arranged to pick a new rubber up at the 2001 National Rally. The rally also allowed me to have a good look at P4 rear windows. So apologies to those of you who were alarmed by some bearded bloke gazing at your car's rear window, I was making an assessment of the shape of the external rubber seal and also the alignment with the surrounding bodywork.

Two things arose from this casual investigation. Firstly, it is definite that the original window rubbers have a more generous external section of rubber providing more rubber for the frame to sink into externally. Secondly, even cars with original fit rear windows do not always have frames that are completely flush with the line of the bodywork all the way around the aperture.

So armed with this added information I fitted the new rubber to the frame. The first thing I noticed was that the profile of the new rubber's section that fits into the frame was thicker and this made fitting the rubber extremely difficult around the corners. However, the additional external profile looked very promising. The moment then came to fit the blighter…

Jason and myself found a convenient date and the battle lines were drawn, this window was going in and that was that! We went through the same process as before and after much joint effort the window was in. The near-side bottom corner was again not flush with the bodywork as it was impossible to achieve this with all other parts of the frame in the correct place. However, with the more generous external profile of the seal this did not matter so much. Unfortunately, the rubber seal was not flush with the bodywork at the nearside bottom corner and required being stuck to the bodywork, this would have not been needed with a more generous external profile of rubber.

Key points for the whole process:

- Roger's article is a priceless guide to the process.
- Make a note of the number of shims required for each window bracket before removing your rear window.
- Do not expect perfect alignment with the surrounding bodywork, adjust the position of the window frame with the clamps and levering until you obtain the best overall fit.
- Somebody needs to manufacture a rear window rubber seal that is a combination of the inner section of the first type I purchased with the outer section of the second type I purchased. I think that a larger section of external rubber to meet with thc external bodywork is required.

  I would like to thank the following people for all their help in what for me was the hardest task I can remember:

- Jean for helping me during the winter months, she was wrapped in several layers of clothes to keep warm in the freezing cold garage, during my initial attempts. She certainly heard some colourful words that until that time she probably didn't think I had in me!

- Roger for his article, the loan of his clamps and the moral support.
- Jason for his help during the final onslaught.

It may be that you would not have encountered the above difficulties but I hope my experience helps you in some way. I would be more than happy to give moral support to anybody considering the battle of their rear window. Please contact Stan Johnstone for my details.

And finally, if you need to do the same job you have my sympathy!

Eamonn Burnell ***2001 Yearbook***

THE OFT FORGOTTEN STEERING RELAY

Of all the parts on the P4 chassis that seem to get the least attention at service time, I think the steering relay (or Idler) unit must be it. How many of you remember to check the oil level, if indeed there is one! Very often the seal has given up trying to keep the oil in and the unit soon becomes empty and hence unlubricated, causing rapid wear in the bushes. The MOT tester seems to enjoy finding excessive play in this unit, but at the same time isn't generally interested in putting the matter right because of the labour involved.

Ideally, to tackle the job thoroughly, the relay will have to be removed from the car. Do not concern yourself with trying to remove the drop-arm from its main spindle, as this very often does not want to be removed unless you possess a very substantial hydraulic puller. Just release the track-rod joints from it and leave it still attached to the unit. The bolts holding the relay unit to the chassis are, generally speaking, very stubborn and will require a good soaking in a releasing agent such as Plus-Gas. The bolts tend to corrode in the aluminium casting of the relay unit body.

Once removed from the chassis, the unit can be stripped of all its components. Start by removing the oil filler plug, which will reveal a circlip just inside the body. After removing the circlip, the shaft can be pressed out of the collar, downwards, out of the bottom of the relay unit. Again the collar can be reluctant to break its fit with the shaft and ideally a press should be used to gently get it on the move. If a press isn't available, a brass drift and a heavy hammer can be employed.

Carefully prise out the seal from the base of the housing. Clean all the parts and then try the shaft back inside the bushes to detect for any wear. Hopefully there will be no perceptible movement, but if there is, the bushes will have to be pressed out and new ones fitted.

If new bushes are fitted, make sure there is absolutely no tightness when the shaft is installed. If there is any interference between shaft and bushes, they will have to be carefully reamed out. The end-float (i.e. up and down movement) is controlled by the shoulder of the top bush and the collar and circlip groove. Ideally you should have .002" / .003" thou. end-float (.05 / .07 mm). If this cannot be achieved you can carry out the modification which is detailed in the Rover Workshop manual. This will require the oil-filler plug to be drilled and tapped ¼" BSF. A good quality high textile 1½" x ¼" BSF hexagon head bolt will have to be used in this modification. A 'soft' bolt will wear very rapidly, so make sure to use a 'Rover' type quality bolt. Do not forget to saw a slot across the hexagon for the bolt to allow oil to enter the idler shaft. Also cut one in the end of the thread to facilitate the use of a screwdriver for adjustment of the end float. Once the relay is bolted back onto the chassis (don't forget to use copper-slip grease on the mounting bolts) and the track-rod ends are attached, the end-float can be adjusted by turning the bolt down on to the shaft unit it just touches and then tighten the lock nut. Use 90 EP oil in the unit which hopefully, after all your labour, will not trickle out past the seal!

This little job should keep you on the straight and narrow providing of course, there is not any other wear existing in the steering box, swivel pins and track-rod ends.

Stan Johnstone ***1994 Yearbook***

THE LOWLY MANIFOLD

The exhaust manifolds of the P4 may seem to most of you, to be a pretty basic piece of Rover engineering to warrant a mention in your Yearbook, but believe me, when I tell you, that there is a lot to be learned about the lowly P4 manifold.

Produced for The Rover Company by Qualcast, a business more famously known for producing lawnmowers, the ignominious exhaust manifold, barely got more than a sentence of mention in the workshop manual, but in fact, has quite a bit of explaining to do for itself!

Fundamentally, there are six types of manifold fitted to the P4 range:– the '75/90', '60', '80', '105', '100/95' and '110'. They were all 1¾" bore, apart from the '105' and '110' which are 2" bore. The '60' has now become the dodo of all of them and is practically extinct, a rarity indeed. Whatever model you run, it makes sense to locate a spare for the future. When they have finally got used (or abused), they will be a difficult and expensive part to re-manufacture.

Theoretically, they should last forever, but like almost everything else on this planet, it gets touched by man and therefore gets maltreated. How could you maltreat an exhaust manifold, I hear you say. Well, you can, quite simply by over tightening. The problem is, once the manifold is removed from the engine block, it releases its stresses of being firstly overtightened and then of course, heated and cooled infinitum, and relaxes, as it were, into a state where the mating faces are out of flatness in relation to each other. If you were to lay the manifold on to a flat surface, you would see quite a few gaps. The only real way to fit a manifold is to machine these faces into a common plane. Once this is done, it can be lightly clamped back on to the engine block. The secret of it all, is in the design of the clamps, which are cleverly designed, so that they only make line contact with the manifold. If you look at the end view of a clamp, you will see that it has a radius built into its base, so that it makes contact along its centre line. I see quite a high proportion of P4's with their clamps actually fitted upside down with the circular bosses, which are of course, the 'land' for the washer and nut to sit squarely on, facing the block.

The '80' manifold is a different set-up entirely, whereby the inlet manifold is sandwiched between the exhaust manifold. These should ideally be bolted onto a 'tooling' block and machined as one, and are certainly not a DIY proposition, but should be done if you want a gas-tight seal.

Once the exhaust manifolds are machined flat, they only need to be lightly bolted to the block, so that on warming up, they are actually allowed to expand underneath the line contact of the clamps and again when the engine is cooling down after use. It is very common to see a manifold cracked because they have been clamped too tight and hence not allowed to move underneath the clamping areas.

So to sum up, if you cannot for one reason or another, have the manifold machined flat, then the only option you have is to overtighten the bolts, so as to get a seal and overcome the distortion, but of course, you then run the risk of cracking the manifold.

Stan Johnstone ***1998 Yearbook***

SLUDGE BUSTER

Not being a daily driver, my cars get their cooling system washout less frequently than Stan recommends. The usual method is to disconnect the rubber elbow at the heater top connection and attach a garden hose instead. With the heater valve fully open, run the hose with the radiator cap off and both engine block and radiator drains open. The heater valve needs to be opened up before the water is turned on, because it won't take kindly to full mains pressure. After a long flush, reverse the flow and give another good flush. This time the heater valve can be closed since water can escape through the drain taps and the radiator top.

The problem with this method is amount of water consumed, especially if you have a metered supply. Since most of the sludge comes out in the first minute or so, turn off the water when the wash water is clear. Reverse the connections as above, but don't flush until the system has fully drained down. The eddy currents created as the wash water purges air from the system has a powerful scouring action and a lot more solid will be shifted. Each time the water runs clear, reverse the connections and allow to drain down before restarting the flush. Even more filth can be shifted by blowing through whilst the water is going in through the heater top connection. You need good lungs for this! The tube to blow down is the one that feeds the top of the heater, i.e., from where the rubber elbow has been removed. The super clean system on my 105 is so efficient that it takes noticeably longer to reach operating temperature. I may have to invest in a radiator muff!

Dave Smith ***March 2002***

GET YOUR PETROL GAUGE WORKING AGAIN

I have spoken to quite a few P4 owners since becoming an owner of a 1962 100 myself eighteen months ago, and it would seem that many owners are blessed with petrol gauges telling us we have the same petrol level in our tanks day in, day out. Many people, myself included, put off remedying this fault and just use the odometer to keep an eye on petrol consumption.

If I'd known just how easy it was to rectify, I would have done so ages ago. The job should take no longer than an hour and requires very little mechanical knowledge. In the majority of cases it is simply the float arm of the petrol tank sender unit seizing. If you don't fancy freeing up the sender unit yourself, you can buy reconditioned ones at the Guild's National Rally. I paid five pounds for one last year and it was perfect. So after freeing mine I now have a spare one just in case.

Below is the procedure I used to repair mine. It is worth noting that when working with petrol you may want to wear some disposable, impervious gloves. I know the content of benzene petrol has been significantly reduced in the last few years, but benzene is still carcinogenic, so take care.

Changing or repairing the tank sender unit:

1. Disconnect the battery lead
2. Jack up offside rear of car, making sure it is properly supported
3. Remove offside rear wheel
4. Remove all wires, located on the sender unit

5. Carefully undo the six setscrews securing the sender unit and slowly withdraw it. (If the tank is about a quarter full there should be no spillage)
6. Be careful not to damage the arm and float
7. Check joint washer for damage and replace if necessary. (part no. 527347)
8. Reverse sequence for refitting sender unit

If on inspection of your sender unit you find that the arm does not move, this is most likely to be the fault. This is simply rectified by the following:

1. Place sender unit loosely in a vice, so the arm is facing downwards
2. Undo four screw securing the cover plate and remove, being careful not to damage rubber seal (I don't think a replacement is available)
3. Spray an electrical lubricant into the sender unit and leave to soak
4. Carefully attempt to move float arm. Do not use excessive force
5. Once the float arm is moving, continue to lubricate and exercise arm
6. Replace cover and clean of external surfaces. The unit is ready to be refitted

You can test the sender unit prior to refitting by measuring the change in resistance as you move the float arm, using a multimeter. Just connect the meter to the same points as the vehicle wiring does.

Jason Dorey ***May 2002***

OH! THAT RING IS LEAKING!

I have been running my Rover 110, incidentally also affectionately called 'The Duchess', since 1997 and one noticeable characteristic has been the lack of need for replacement coolant. Last year, however, two consecutive runs to the Steam Rally at Carrog on the Llangollen Railway, about 160 miles in one weekend, resulted in about two pints of water being needed to bring the radiator to normal. No immediately obvious leakage could be seen with the engine warm. At the next meeting of the Cheshire and North Wales P4 group the pundits were also at a loss as to the means of escape. Could it be that the radiator cap was weak and allowing water out under the pressure of running conditions? It might be the dreaded cylinder head gasket allowing water into the cylinders, but there was no evidence of the tell-tale white exhaust smoke. Even worse, water could be finding its way into the oil. A check of the oil level showed no alarming increase, and the oil looked reasonably clean.

The answer to my suspicions was confirmed by Geoff Vaughan, who along with John Cumberlidge act as representatives for our group. Geoff looked at, and then touched, the thermostat bypass O-ring above the water pump which, it must be said, looked rather contaminated. On interfering with the Duchess's intimate parts she'd wee'd up his arm!! Close inspection of the water pump body revealed a pale antifreeze stain where coolant had escaped but rapidly evaporated on the hot aluminium. Geoff, in his usual quiet manner, had effected a diagnosis of the problem. All that remained was to replace the offending O-ring. It may be prudent, at this point, to explain what this bypass business is all about. When the engine and therefore the cooling water is cold, the thermostat remains closed so the water in the engine block can warm up quickly. In order to prevent excessive pressure build-up by the water pump, a small bypass is arranged around the thermostat. This is usually a short hose between the body pump body and the underside of the thermostat housing; anyone familiar with the BMC A series engine will be well aware of the problem of changing this component. Our P4 engines have a similar arrangement until we come to the Westlake cylinder head fitted to the 110 model. I cannot say precisely why, but the 110 has a machined surface on the water pump and a machined surface on the underside of the thermostat housing with a small hole, in alignment, in each. When the pump is in place an O-ring is trapped between the two and forms the seal. Now much conjecture was generated by the proposed replacement of this O-ring and the worst-case scenario was the cylinder head would have to be removed. This is NOT the case.

The next perceived difficulty suggested was that the nine bolts holding the water pump in place would be seized and some would shear off when an attempt was made to remove them. Fortunately in my case this did not happen either, but it is a real possibility because of corrosion caused by the effect of water on the dissimilar metals, i.e. aluminium pump body and steel bolts.

I decided that I should undertake this repair, and a telephone call to John Wearing confirmed that the head need not come off. This was followed by an O-ring, gasket and a set of UNF bolts, just in case, by return of post. I felt that room to work would be essential, and decided to remove the bonnet lock, grille support brackets and the grille. This is not as difficult as it first looks, but the two bottom radiator bolts are a little confined. Remember to drain the radiator and engine block, then remove the top and bottom hoses. The clip on the bottom hose can be difficult to get at if the jubilee clip screw is not visible on top. The radiator is held by two studs projecting below into the support channel and three setscrews each side, captive in the vertical angle and attached to the inner wings. Before the radiator can be lifted out, the tie bar, which on the 110 carries a cowl over the fan, must also be removed. Access can now be gained to the dynamo brackets and fan belt adjustment which can be slackened off. At this point it's worth considering removing the dynamo for routine maintenance. Once the fan belt has been removed, the four setscrews holding the large four bladed fan in place can be removed. I marked the fan so that it could be replaced in the same position, this was followed by the steel inlet pipe to the pump and its rubber connection.

We now come to the main event, so to speak, the nine set bolts retaining the water pump have four different lengths between them, so it is advisable to note the position of each one, although replacement becomes fairly obvious. In my case all the bolts came out easily without shearing. If excessive tightness is encountered, gentle persuasion by tapping with a light hammer may help. The pump body can then be prized carefully off the engine block. Again, some persuasion with a block of wood may help to loosen the gasket, the pump is located by two dowels which could also be corroded. The offending O-ring virtually fell apart in a sticky mess, although the old gasket must be removed and the mating services cleaned carefully of all remaining gasket and oil or grease. It is also a good time to check the water pump bearing and seal, to see that there is no evidence of leakage or wear in the bearing, otherwise a replacement may be needed.

Refitting of the pump and seal is, as Stan would say, a simple case of 'Sling it all back together in reverse order.' However there are one or two points to watch for. The cause of all the bother – the O-ring – is positioned over a small spigot on the pump. Check that the hole is clear and apply a touch of jointing compound to keep it in place. I found that if the bolts are placed in the pump body, the gasket can be located taking care to make sure that it fits over the location dowels in the block. If the dowels have left the block it is easier to put them back in place to aid location. As the pump is offered up to the underside of the cylinder head to slightly compress the O-ring, the dowels can be located, then the set screws engaged carefully into their respective holes and tightened. No order or torque setting is given to these bolts but common sense and 'easy does it' seems to be the order of the day. I used a proprietary thread locking adhesive on both pump and fan screws as an extra precaution. When replacing the pulley and fan it should be noted that there is a spacer for the fan blade which is fitted in front of the fan on later models. The fan belt, preferably a new one, can easily be fitted and tensioned at this point with the radiator still out. The recommended adjustment for a 110 is such that the belt can be depressed ¼ to ½ and inch (6.5 to 9.5mm if you're in Brussels!) but I prefer to err on the slack side to prevent too much load on the dynamo and water pump bearings. With the radiator safely located the hoses can be replaced, the bottom hose can be a little difficult in that the jubilee clip will turn on the hose when you try to tighten it. I found that if it is pre-tightened just enough to grip the pipe with a screw uppermost. Then it can be screwed up as required, taking great care not to damage the radiator core. Having had to drain the antifreeze, this is also a good time to flush out the system with clean water and add new antifreeze. The system holds 24 pints (13.9litres), so quite a lot of antifreeze will be required. It will probably pay to buy a gallon and add six pints to give the required 25% strength to protect against corrosion rather than frost (we all keep our P4s in heated garages, don't we?) Replace the radiator pressure cap, run the engine up to temperature and look for leaks with the heater valve open, then check the water level and top up as required. Do not overfill. This can be done before the grille support brackets and grille are replaced, but do not drive the car as there is no means of holding the bonnet down! Finally, refit the grille and bonnet locking mechanism. Note that there is a small hole in the grille top support to the left of the badge to give access through the grille, with a thin screwdriver, to release the bonnet lock if the cable becomes loose.

I hope that this tale will encourage owners to try their own maintenance and keep our wonderful P4s running.

Trevor Vernon ***May 2002***

CONVERSION OF A P4 TO NEGATIVE EARTH WIRING

My 1964 model 95 was never fitted with a radio when I bought it about three years ago, and I decided to install an up-to-date four speaker radio and single slot CD player. Not authentic I know, but they sound so good in my modern car, and prices have dropped so much that a really excellent model is available for around £150 plus speakers. After I brought it home from Halfords, I read the instructions and realised that it was only for use in negative earth configuration, and that no modern radios were available for fitting to a positive earth car.

A quick phone call to Stan revealed that it was quite simple to switch the Rover from positive to negative earth (or so it seemed at the time). Fitting the speakers was fairly easy; two large oval speakers fitted in the front footwell kick panels and two smaller units were mounted in the rear shelf behind the back seat. I bought a second-hand dashboard centre panel at the Rover Spares Day and substituted it for the existing wooden fold-down panel. This has a radio slot cut out (amazingly this aperture is the same standard size as used today, 40 years later) but I discarded the single tiny speaker fitted above it.

Then came the interesting bit; according to Stan all I had to do was to reverse the battery terminals, then restart the engine and the dynamo would re-polarise itself and all would be complete for wiring in the radio. I changed the battery terminal leads over with trepidation and a fire extinguisher to hand, but nothing untoward occurred. Then I switched on the ignition and turned the starter motor. The engine caught and fired; amazingly it worked! My elation turned to gloom a minute or so later, when the engine died and steadfastly refused to start. I checked for sparks and all seemed fine, so it looked like a lack of fuel. Sure enough, on disconnecting the feed pipe from the carb, there was no fuel being pumped by either of the two pumps. I removed the twin pumps and immediately noticed, on undoing the covers, evidence of burning on the wiring to the contacts. Also there was a large red sticker on the outside of the pump proclaiming '12 volts positive earth only'; the damage had been caused by the arcing of two small diodes wired across thc points, the double entry SU pump used on the 95 and 100 being different to the pump used on the earlier 80 and 100 models. I suspect these diodes were intended to cut down arcing on the points and/or to help with radio suppression. The points on the pumps were badly damaged too and needed replacement. A friend who is an electronics engineer helped me by supplying new diodes that were installed in the reverse to original. After a careful rebuild, both pumps were bench tested and found to work again. Success! The pumps were reinstalled and everything put back in place and the car started and fired up reluctantly, but it ran.

A quick phone call to Stan elucidated a further suggestion that I should have switched the polarity of the coil, the cause of reluctant starting. Apparently the spark voltage is reduced some 40% if not, and sure enough, starting from then on was right on the button! The installation of the new stereo was completed within a day and I now had a sound system to be proud of. Subsequently, the car underwent a major bare metal body restoration at a small bodyshop in Harpenden. Foolishly, I told the owner that I was not in a hurry and to save costs, he could do the work when his other commercial work allowed. As layers of paint were removed and 40 years of corrosion and accident damage revealed, it was clear new aluminium doors purchased from Richard Hedger would simplify the repair. Whenever I visited bodyshop, the car was 'a few weeks away from completion'... That meant it was two years later before I received the car back with its new, immaculate two-tone green exterior.

Shortly after that I found a car upholsterer in Baldock, who covered the front seats in new green leather for a bargain price. This time I only had to wait two or three months to complete the restoration.

The last problem to be resolved subsequent to the polarity conversion was that the fuel gauge and temperature gauge no longer functioned. These items are fed a constant 10 Volts by a rather bizarre stabiliser mounted behind the dashboard consisting of a bi-metallic strip. Swapping the polarity of this device did not cure the gauges' functionality so I removed this device and connected them direct via the 12 Volt supply. Sure enough they now functioned but clearly were not reading right as the temperature was always in the red zone after warm-up. A quick conversation with my friend Mr Electron and he made up a voltage drop circuit consisting of a 10 Volt Zener diode and ballast resistor and sure enough both gauges read normal. The fuel gauge was even accurate as it showed 'E' on the drive home, and for one heart-stopping moment the engine died as I feverishly switched to the reserve pump. A few seconds later it caught and we proceeded.

Sean Coles ***September 2002***

MISCELLANEOUS RESTORATION ADVICE

I've written here about things which might have gone smoother during the ten-year restoration of 500 CMU, my 1954 90, if I'd done it the right way. I suppose these observations won't be particularly helpful unless they're read by someone who is restoring their P4, as most of these jobs wouldn't be considered to be routine ones. 500 CMU is up and running now, although a few things will still need tidying here and there, especially where I replaced a part nine or ten years ago and it now needs doing again, even though it might never have been used!

Heater box site control levers and the rubber drain tube for the fresh air intake

Whilst trying to remove the heater control levers on the side of the heater box so I could have them plated, I accidentally broke the support bracket of the lever nearest to the bulkhead (when in situ). The heater box was not yet reinstalled in the car so access was easy, but these two small levers are a very tight fit on the swivelling rods within the box, and they must be removed to allow the support bracket to come off. I removed the heater box from another car so I could try again, only this time the levers were so tight on their rods that they just wouldn't let go. I decided to apply heat to the levers to free them, and this worked a treat – at least for the one nearest to the heater tap on the outside of the box. This lever came off easily, and once this was off the support bracket was easily released by undoing the two fastening screws. But it was the other bracket I had broken, so I got to work on that one. Well, I got a surprise when I applied heat to the lever only to see the support bracket quickly melt and distort. So that was that, two of them I'd ruined! The larger one is made of steel, but the smaller one is a special shape and is thicker at the far end where it is tapped for the control wire clamp screw. This special bracket is made out of some sort of soft alloy and is evidently very easily damaged. Watch out for this if you ever try to remove it! It is really worth the extra care needed to preserve this part undamaged (or make an equivalent part out of steel?). The bracket is not accessible unless the whole box is first removed, and this only after taking off the blower unit first. Once this much has been removed it is a good time to check, and if necessary replace, part number 312666 – rubber drain tube air duct. This tube cannot be accessed without removing the blower unit, and it's a bit of a game replacing it even then! I didn't even know it was supposed to be there and I noticed the diagram and reference to it in the Parts Book. On 500 CMU it was missing and had never been there during my thirteen years of ownership. I felt about with a hand inside the large hole normally occupied by the dash-to-blower connecting sleeve and found remnants of the original disintegrated tube! Roger Dealtry, in his website www.head2head.free-online.co.uk/Rover/spares.htm tells us that Edgware Motor Accessories, tel. 0208 943 4789/9311 is the only place he knows of who stock the original type rubber tube for this application. To my knowledge also he's right, but Edgware themselves have no record of the size and length required for any particular make of vehicle, or which might be suitable for the Rover P4, so you'll have to tell them exactly what you want if you try to order one. However once ordered they offer a very good next day delivery service. I asked for one metre of thin rubber tubing with half inch internal diameter, which is just about right.

Quarter-light window seals

The official P4 Workshop Manual is very economical when it comes to describing the side vent glass and its removal. Nothing is said in this book about the seals therein. Therefore it will be of great benefit to anyone thinking about changing these seals to obtain a copy of Carl Haworth's booklet on changing them and thoroughly read this first (now a collector's item, I imagine – I got my booklet and seals when they were first remanufactured, about ten years ago). It makes jolly good reading anyway, and you may be surprised by the large number of small parts you will need to do this job properly. (There are eight rivets – some of different sizes – needed for each unit – nine should you disassemble the swivel hinge – and several seal locating pins which pass through and secure each seal). I was lucky enough to have a couple of spare quarter light windows to play with before attempting the real job. Not really necessary but very useful if you are a bit of a muppet like me. As Carl says in his book: it is all much clearer when the parts are in front of you, but this job is tricky if you haven't seen a quarter light window out of the car and examined its component parts.

If you are replacing the quarter seals it will make a great difference to the end result if you first have the chrome window frames re-plated. This can only be done if the whole unit is disassembled. The glass will come out easily enough if you persuade it gently. I started to remove it with a small, blunt screwdriver in one end between the frame and the glass, and with the whole thing held by the glass in between the jaws of a vice. (The jaws must not be over tightened and must be protected, and do not force. I used an old towel in between them but it was risky as the glass must not be strained. Carl Haworth rightfully suggests the use of fibre jaw inserts in the vice to act as protectors. This way I could gently lever the seal frame away from the glass. The delicate weather seal on the unseen glass edge is particularly fragile, so take care with it and do your utmost to preserve it. I didn't break any glass doing the job this way until I dropped one frame later on – after the job was finished, and then had to rely on my spares for another pane of glass!

Brake pipes

If you intend to replace the brake pipes on your P4 you will want to think twice about the one right at the front, unless you are removing the engine. It is even easier if the wings are also removed. This pipe is very difficult indeed to get at and secure with the correct chassis clips unless you have easy access with at least no engine being in the way. Of course I didn't do this the easy way. The front wings were already fitted, and I had put the engine back in too! The job just takes far longer this way and is so fiddly, as you are trying to work upwards with almost inaccessible spaces for your hands. The chassis curves awkwardly too. Make life easier for yourself by remembering to replace your front brake pipe whilst the engine is out or the front wings off!

Wiring loom

Another point I got the wrong way round, although it is not a total disaster, as it is quite easy (although laborious) to correct – was my not realising that the wiring loom must pass inside the A posts, for want of a better description; not in front of them – as mine were until I then tried to secure the interior windscreen surround which of course won't fit like this. There is room for the loom behind part numbers 310531 and 310532 – spacer, windshield opening LH/RH, although it might not seem like it at first – it has to be neatly tucked back. The upper front door rubbers will have to be removed again if you get this one wrong, as you will have to remove the 'hidden' bolts which secure these two parts.

Distributor housing

If you are in the habit of lifting your engine the way I did – by attaching the hoist in some way around the distributor housing you may well be sorry when you try to reinstall the distributor. It may not look any different if you have lifted your engine with it, but believe me, this part is made of alloy and can easily get distorted. And then the distributor won't go in. And then you'll probably say what I said.

Rear bumper support brackets/petrol filler tube

Ideally, the rear bumper support brackets must be replaced before the petrol tank goes in, assuming you have previously removed the tank. You will need the free space to tighten the largest bumper support brackets nuts and bolts. The rubber connector, tank to filler (part number 235168) and the rubber tube for the air vent pipe (part number 235815) are both difficult to get back in correctly. The thicker pipe depends on the filler neck being accessible so it can be attached and tightened, and the thinner one passes through the cover for the petrol filler tube (part number 312036) which cannot be tightened down too early as you might need to be flexible when trying to tighten the screws of the clips below it which belong to the aforementioned pipe. To make it all worse, access is so limited here and you'll have to find your own way of doing it. These pipes should first be placed correctly, one end on the tank and the other to attach to the filler tube for the large pipe, and the filler neck side extension tube to the fuel tank extension tube, passing through the cover plate en route for longer, narrower one. They are both swines to get in place if the tank and filler tube have been fully tightened down, so leave things loose until the last moment here before fully tightening. This is one of those jobs which will become obvious when you are lying there on your back trying to get it all back in the right order, but once again you'll be sorry if you have to undo everything and start again (like I did!). It all sounds horribly complicated, but you'll understand it if you study the parts before you try to assemble them.

Foresight

I cannot emphasise enough how relieved I was throughout the building process that I had taken the trouble to label everything carefully and place the parts neatly into bags when I took the car apart ten years ago. I would simply not have been able to remember what was what, apart from the very obvious things. I would advise you to take the time to make drawings in your parts book too as these will help, together with notes of any modifications you find or make. Even broken or worn out parts may serve well as a pattern, so don't throw anything away until you are doubly sure you don't need it any more. Even then, think of your fellow club members – you may not need that old part if you now have a better one, but someone else might be glad of it.

Mark Kelly ***September 2002***

IRREGULAR CURRENT = BLACK SMOKE

I found Dave Smith's article on Auntie's ignition and charging system in the May Overdrive very interesting, as I experienced exactly the same fault but with much more serious damage and consequences.

My problem began back in January this year when I collected the car from the shipping agent in London after Auntie's return from Thailand. Being transported from a warm climate of 30 degrees C down minus 6 degrees C in four weeks did the electrical system no good, resulting in a flat battery.

After cranking the engine over for ten minutes, assisted by polar bears and penguins, 'Auntie' still refused to start so I borrowed a powerpack booster gadget. She started reluctantly, but was okay once she'd warmed up, which took a fair while on such a chilly day. I drove her home nearly 200 miles to Yorkshire without any trouble, except for the last ten miles, when darkness fell and I switched on the headlights. I had just come off the A1 at Doncaster and the main road from there into Barnsley is mostly unlit, so I was alarmed to find the headlamps on one candlepower between them! I managed to get home, put 'Auntie' into the garage, disconnect the battery and laid her up for three months while I jetted off to the sun.

On my return in April I checked everything over and found the fan belt nearly in two. I put the battery on charge, but after two days it became apparent that it was unserviceable and not holding its charge. Once a new high-power battery and a new fan belt were fitted she started immediately, with the ammeter showing positive charge and the old red light behaving itself.

Still unsure of curing the problem, I proceeded to clean all the main contacts on the ignition circuit and double checked everything. I took the cover off the regulator box and gently cleaned the bottom spring contact which was pitted and refitted the cover. Again the ammeter was functioning normally, but I decided to go to an auto electrician to have everything checked, as the ignition light was coming on and off at will, regardless of engine speed.

Malcolm, of Brook Electrical, proceeded to check everything with his voltmeter and found the regulator box to have an erratic operation. He removed it and put it on his test bench to set it all up and found it to be way out of 'tune'. The previous owner had in fact tweaked the dynamo output up to absolute maximum, hence the cooked battery and the voltage regulator cut-outs were working now and again. Malcolm re-set everything and refitted it to the car and all seemed okay again. He advised me to consider fitting a new regulator, as the original was very old and couldn't be relied on as all settings had been tampered with. The regulator box contains three magnetic coils, which are all temperature compensated in operation – a very delicate piece of equipment indeed.

I decided to take Malcolm's advice and as a visit to John Waring was only a few days away, I thought I'd collect and fit it whilst I was there. The car ran okay for the next couple of days and a spell of dry weather prompted me to clean the glass petrol filter bowl and fit a second in-line plastic one, plus clean plugs, etc. The car was at the top of the drive, standing idle, ignition off, while I was busy cleaning the filter in the garage.

I glanced towards Auntie and saw some wisps of black smoke emitting from under the bonnet. A Basil Fawlty like dash ensued, where I discovered the smoke was coming from the regulator box and adjoining wiring. It took me around 90 seconds to remove the back seat and the battery cover to uncouple the battery, but it had burned all the installation and fabric covering on approximately 20 wires from the fuse box down to the bell housing, and onto the dynamo and HT coil. Gulp!

It could have been much worse. The old Rover could have caught fire and I was extremely lucky to have stopped it in time.

I ordered a new regulator box straight away, which arrived the next morning (service with a smile as usual from John Waring, bless him) and spent a whole day carefully removing all the old wires one at a time and clipping them up. I then fitted the new regulator box and recoupled the battery, which I'd recharged, switched on and............ nothing happened! I was sure I'd got all the wires coupled up correctly. So I checked out the back of the ammeter. On removal of the dash panel I found, to my horror, the ammeter was well and truly 'cooked' as well, but luckily there was no damage to the surrounding wiring. Bridging the ammeter terminals restored the ignition circuit and Auntie breathed again, but now the red ignition warning light remained on and refused to go out at any engine speed.

Back to Malcolm, the friendly auto electrician, who gave me 8/10 for the rewiring (not bad for a mechanical engineer) to check everything. He found the dynamo was not giving any charge so he tried to 'kick start' it with the supply wires to the regulator box, but to no avail. He suspected either an insulation failure or internal failure in the unit as a result of the fire damage.

I recharged the battery that evening and drove to John Waring's – 57 miles away – then next day fitted a new dynamo and ammeter, and with fingers crossed Auntie is behaving herself again. Now I've got some mechanical work to do, which should be a cinch after this saga.

Advice to Auntie owners:

1. Remember the electrics on Auntie are very old. At the first sign of trouble replace the offending unit and have the system checked by qualified auto electrician who understands classic car electrical systems, not a modern whizz kid
2. Do not in any circumstances tinker with the adjustments on voltage regulator
3. A new regulator is set on assembly and should be returned if faulty, as it is guaranteed
4. Do not attempt to fit a 90 regulator to a 100 – they are different. Ask for the correct one
5. Fit an isolator switch to the ignition circuit which is easy to get at, to minimise damage

By the way, the insulation on the top coil of the old regulator had failed, causing a massive short-circuit, hence the ensuing damage. Many thanks again to John Waring for all his help, advice and service, cups of tea and who even found time to show us his fabulous Marauder, mint Land Rover and his latest 100 project.

Tony Brown ***September 2002***

STICKING OVERDRIVE

Setting out for work last winter, I thought I felt the familiar surge as the overdrive cut in. Funny, I thought, I haven't selected overdrive yet. At this point I should explain that I have bypassed all the 105R circuitry in favour of a switch from a manual car. Overdrive is thus available on both emergency low and normal drive. I haven't yet interlocked this to the reverse gear and so it is possible to select overdrive in reverse.

Reverse and overdrive together usually result in a wrecked overdrive unit and so I automatically check the overdrive switch before engaging reverse.

On the rare occasions when I have been distracted, it is soon obvious that something is amiss, since on reasonably level surfaces the tickover revs have enough torque for the car to move freely. With overdrive engaged, the car locks up in reverse, but seems to free up without apparent damage as soon as overdrive is deselected, provided that the engine hasn't been revved to create enough torque to destroy the unit.

Continuing my journey to work, all was well until I tried to reverse into my usual parking space – yes, overdrive was still engaged. Nothing for it but to circle the car park until I could find an empty pair of through bays in order to go in and out without reversing. I had visions of driving into a dead bay and not being able to reverse out!

Working twelve hour shifts in winter means cold and darkness at either end of the shift, neither of which is conducive to crawling under cars. I initially thought that my rewiring was suspect and so disconnected the supply to the overdrive relay; all that could be done from under the bonnet. Unfortunately this didn't improve things, so it was time to consult the Workshop Manual.

In the meantime, journeys (well, more the parking) had to be planned so as to avoid reverse gear. Quite a few times I had to resort to putting a front wheel on the pavement and even so ended up being too close to the car parked in front. That studying of the Workshop Manual revealed that the overdrive doesn't cut in until enough oil pressure has accumulated and therefore it will disengage a little while after parking up. Sure enough, a walk to the newsagents gave time for pressure to fall enough to engage reverse and set back from the car in front. If your overdrive should stick, I reckon you have about twenty seconds from cold in which to reverse before the overdrive tries to cut in. My fears of being stranded in a dead-end bay were, therefore, unfounded, but I still didn't take the risk!.

At the time, I had heavy work commitments and so had to defer repairs, but in any case, the Workshop Manual didn't help pinpoint the cause of the problem. As the failure had occurred shortly after I had topped up the gearbox oil, I thought that maybe a foreign body had got in there and so I changed the oil and was very careful about cleaning everything in petrol before refilling with fresh oil. Still no joy.

In the end I decided that the transmission tunnel cover would have to come off to give a clear view of what was or wasn't happening. Easier said than done. Removing the seat and carpets is no problem but the dozens of screws holding the cover on can be reluctant to let go. The problem is that the thread is longer than the floorpan is thick, and so the portion that protrudes below the car is subject to 40-odd years of corrosion (except for the bits that get covered in self changing engine oil). Most screws will succumb to a large screwdriver – the kind with a hexagonal portion at the top of the shank, allowing a spanner to apply greater torque. Even so, full body weight is required on the handle to stop the screwdriver from lifting out of the screw slot. Despite the above and liberal doses of easing oil, a few screws ended up somewhat gnarled and could only be extracted with mole grips.

With the electrical connections reinstated, it was now easy to deduce that the operating shaft was seized solid in the 'on' position. How not to do it is what I did next.

I removed the solenoid, sprayed easing oil over the works and moved the operating level to and fro until it loosened up. Wrong. All I had done was work the lever loose on the shaft!

At the opposite end of the shaft is a setting lever which is firmly located on the hexagonal end of the shaft. Waggling this repeatedly did the trick. All that remained was to reassemble as per the Workshop Manual. As I don't have the proper Rover setting bar, I had to find something else to line up the hole in the setting lever with that in the casing. I used an Allen key – not the usual hexagonal one, but one with a Torx bit, since these are closer to being circular. Another idea would have been to use a chopped down drill bit. As I didn't fancy refitting the cover, carpets and seats until I knew that all was okay, I jacked the rear wheels clear of the ground and placed the car on axle stands. With the car in gear and engine running, the indicated speed can be raised to 30 mph. The speedometer will display a satisfying leap as overdrive is operated.

Overdrive then behaved perfectly until January this year when, heading for the M4, it failed to operate. Yes, it would fail during a journey to Pembrokeshire – as far west as one can go without a snorkel. Not having time to fix it (arrival at our host's was going to be after 11.00pm in any case), there was nothing for it but to put up with a noisier journey with poor fuel economy. Without overdrive, the car sounds terribly noisy around 50mph and then wind noise starts to take over until, at 70mph, only the extra accelerator depression gives the game away. Out of curiosity, I measured the fuel consumption by brimming the tank and at 24mpg I thought it wasn't doing badly.

Back home, the failure was traced to a dead relay. I'm still running on a modern Halfords relay piggy-backed onto the original. It looks a mess but with the bonnet shut, who is to know?

Dave Smith ***July 2003***

MORE CROSSED WIRES

I was surprised that nobody responded to the 'Don't get your wires crossed' article in the July 2003 Overdrive, especially the comments from Stan Johnstone.

There have been many articles about hot starting over the years. I had a 100 in the 1990s (the ex-John Burnapp SPL 390) and suffered this problem without any cure. A friend of mine also has 100 that is affected the same way, but my recently acquired 1956 90 isn't affected, either on long, short or hot day journeys. As I never had problems in the 1960s with our 90, 105 or 3-Litre cars, I always felt there had to be a simple explanation; the most probable being fuel vaporisation with modern, volatile fuels. However I think Stan might have the answer.

My current 90 has the coil wired up correctly, and I suspect our Rovers of the 1960s were also correct for positive earth cars. My friend's 100 was incorrectly wired up, so he changed the wires around. As he was recovering from a heart bypass operation, he couldn't road test the car, but he ran it in the garage at different temperatures and for different periods, including a time with a blanket over the radiator, and it would now appear to be cured. Our local Lucas agent said that a coil will work when reversed, but it will give out only a third of the power.

The theory I have is that positive earth cars have not been around for 30 years and, quite understandably, when a coil is reconnected or replaced, it could be wired up in the now 'normal' way, i.e. for negative earth. Also, some coils are actually marked CB and SW but if, as I think, there are negative and positive earth coils and most coils sold today are negative, it is logical to put CB to CB and SW to SW, but this could be wrong.

The only flaw in this theory is that they always start first from cold, even with a third of the coil output. Perhaps the extra strong mixture on choke helps to compensate and, as they are usually fitted with 'performance' coils, the output is higher. But when the car is hot, the vaporised fuel takes more igniting and the weaker spark just isn't enough.

If you look in the Workshop Manual wiring diagrams, there is a section through a coil showing the wirings/windings and you can see that if it is wired up the wrong way round, it will bypass a lot of the internal wiring.

I hope this is the answer. And once again we have Stan to thank. It is worth a try or at least a check.

Colin Gray ***November 2003***

REWIRING

About two years I purchased from Autosparks a new wiring harness, for my 1964 110 with the intention of rewiring the car. I read the article in Overdrive January 1988, which I have to confess frightened me. Two of the photographs showed the 'Guild Auto electrician' inside the engine bay with the engine removed. I think, having read this article, and seeing the photographs, I decided to put off the job until I could pluck up the courage to tackle it!

The harness comes in three pieces, the first from the instrument panel to the inside of the engine bay, the second from the gearbox/handbrake to inside the engine bay and the third from behind either side of the dashboard to the rear of the car via the roof lining.

One morning at the beginning of July I decided to take the bull by the horns, so I went out to my car, lifted the bonnet and with a pair of garden secateurs cut through the harness in two places. I now had to do the job!

I first removed the glove box door and the radio panel. Next to remove was the instrument panel; this is easier than you think, as most of the switches can be removed without disconnecting the wires. Any other wires should be identified by some sort of label.

The next thing I did was to cut through the main harness on the left and right hand side just as it comes through the engine bay, under the wiper motor on the nearside and under the voltage regulator on the offside. The two legs of the harness can now be removed from the left and right hand side gutters along the wings. Don't disconnect the connectors to various components, cut the wires about 3 inches from each connector so that you can easily identify the colours. Now remove the main harness from inside the dashboard but leaving it connected to the instruments panel switches. This looks like a real rat's nest but don't worry.

Now feed the new harness through the scuttle into the engine bay via the grommets in the scuttle. Connect the wires to the fuse box, overdrive relay, starter solenoid and head/side/flasher lights removing the old connectors as you go along. Repeat the offside connections to the control box, horn, dynamo, sump and oil pressure switch then the lights. Again removing the old connectors. Where the harness is clipped to the wing gutters I wrapped a couple of turns of tape around, just to protect the harness from the clips.

The next job is to transfer all the switches from the old harness to the new and connect the instruments, any other wires can also be reconnected, i.e.

those leading to the rear of the car at the door pillar. And while I had the instrument panel off I had it re-sprayed and new lettering added to identify the switches. These can be obtained from 'The Image Works (0115) 944 3111. It now looks a treat.

The small loom from the gearbox/handbrake is the easiest to fit except that to save going under the car, the gearbox tunnel has to be removed.

I have say that the job was far easier than I imagined, so for anyone who is contemplating tackling the job, go for it! It's not as bad as it looks.

Problems encountered

There were problems however. Firstly the wrong size spade was fitted for the starter relay. Incorrect connectors were fitted to the earth wires on the wing gutters. On my 110 the wires to the heater rheostat and choke warning light switch leave the main harness in two sub-forms, whereas on the new harness there was only one sub-form containing both. On some 110s the two controls are on the right side of the heater controls, so this would not have been a problem. On my P4 the heater switch and choke light switch are on either side of the heater controls. However a visit to Autosparks (they are based only four miles from my home) soon sorted out all my problems. They pointed out that that the drawings they used may well have had these connections shown.

Colin Talkes ***November 2003***

(Editor's note: before removing old wires from hollow sections such as the A pillars or the horn wire from inside the hollow steering column, don't forget to attach a piece of string. Leave the string in place to tie to the new wire and pull it through.)

EARLY CAR REAR REFLECTORS

I took my 1954 90 for a run one Sunday morning a few weeks ago, prior to an oil change. After about twenty minutes I returned back home and turned the engine off before jacking the front end. Walking around the rear of the car before going into my garage for new oil, I noticed one of the small red reflectors and chrome retaining rings was missing. So I left the oil draining from the 90 whilst searching the garage floor; no luck. I took my modern car on a trip to see if I could find where the reflector and chrome ring had fallen. I couldn't find them. Of course, they may possibly have dropped off a while ago and I hadn't noticed. These items are an MoT requirement – at least on early P4s – later models have integral reflectors in the rear lamp clusters. I looked around for a replacement and hunted the spares catalogues of our spares suppliers. Nobody listed them, and I was told by one supplier it would be a difficult item to find.

The Internet is a wondrous thing, and eventually I tracked down what I was looking for, at www.vintagesupplies.com at £4.50, although postage etc. doubled the checkout price.

The item I received was just about right so I was happy. I went out put it on the car and then realised why the original reflector and ring had fallen out; the rubber insert where the reflector fits in a bracket below the rear lamps was hardened and weak. A new rubber one was included in my purchase, but this was a screw on type, flat backed, not push fit as per original, with a push in flange. So I made a largish washer to be positioned behind the bracket for the screw in rubber retainer to be screwed to. A retainer to retain the retainer!

Now of course I am looking carefully at the other side and may do the same thing with that one too. Have a look at yours if you have an early car.

Mark Kelly ***January 2004***

PILLAR SWITCHES

David Tame reported problems with his pillar switches for the rear courtesy lights of his 105R. These are often a problem and I can remember them needing a clean even when P4s were only six or seven years old.

The switch, like the one in the door opening, only makes and breaks the black earth wire, so if David disconnects the switch and touches the two ends of the wire together, he should get the rear interior lights to come on (with the rear doors closed, of course). If they don't work, he has a wiring problem or a faulty light fitting, but they probably will work and this indicates the switch is the problem.

When I bought my current P4 last year, the previous owner actually asked me what the switches were for! Needless to say, neither of them worked and I needed to check them.

When I tried to get them out, one of the two connectors was corroded solid and I couldn't undo the screw. I cut the wire off and drilled out the wire core that was left in the socket. Then I tinned the drilled out and clean brass cavity and soldered in a short length of wire, which I then connected to the end of the wire coming out of the door pillar.

I also cleaned up the brass roller and contacts, followed by a smear of Vaseline to stop future corrosion. They now work, although you sometimes have to 'rock' the switch to make good contact. I think cleaning is the answer, but it is difficult to clean the whole of the roller contacts and I also wonder whether, as they wear and/or are cleaned, they are not in such heavy contact with the fixed contacts.

I think the later switches, i.e. Rover 100 etc, were different, so the above may not apply.

Colin Gray ***March 2004***

THE ENGINE: CAUTIONARY TALES AND IMPROVING TALES

Note: some of these remarks apply solely to the 7-bearing, 1959-on engine, which first appeared in the 3-Litre in 1958

If you don't know history of your 7-bearing engine be on your guard – there is every possibility that it's a 'bitsa', i.e. bits of this and bits of that.

The 7-bearing 6-cylinder block introduced in the Rover 3-Litre in 1958, and in 1959 in the Rover 100 as a 2625cc engine, served in saloon cars for only nine years until the demise of the 3-Litre: for another thirteen it soldiered on, again in 2625cc form, in the Land-Rover, and whilst parts may be interchangeable, some will not be identical and indeed may have different characteristics, which may have an impact on other aspects of the engine.

When the 6-cylinder engine was first installed in the Land-Rover IIa in 1967 the compression ratio was reduced, in some cases to as low as 7:1, for use in the Third World, and camshaft timings were adjusted. The engine was rated at 83bhp at 4500 rpm with torque of 128lb/ft at 1500 rpm. A high-capacity oil bath air filter was fitted. The first few engines were fitted with SU carburettors, but later ones used Zenith and finally Stromberg carburettors, the latest examples being tamper-proof; from the engine's first installation in the Land-Rover the cam follower rollers first introduced in 1958 were used in the valve gear, but over the years they were subjected to modification. By 1980 it had to comply with emissions controls which were not even in contemplation in 1958: amongst other innovations, for example, there was an air pump driven from the fan pulley which injected air into the exhaust manifold, and the crankcase was vented into the inlet system. By the end of production, the fan was a plastic moulding and was viscously coupled. Unless you know for certain that your engine is original, take absolutely nothing for granted when rebuilding it.

Check the flywheel – is it the right one for your engine? In the case of the P4 separate flywheels, differently marked, were produced for the 100 and for the 95/110 and are listed accordingly.

Check the camshaft – camshafts for the Rover 100 were marked '100' and Weslake camshafts produced whilst the 95/110 saloons were still in production were marked 'WES'. If you have an unmarked camshaft you need to ascertain what it is. Unless the flywheel and camshaft match, the mark for EP (Exhaust Peak) will indicate at the wrong place.

To identify items when in doubt, you need to consult not only the Workshop Manual for the later P4 saloons but also one for the Land-Rover Series IIa, the 3-Litre saloon and the Land-Rover Series III.

You need to know whether your engine number has the suffix A or B – the Manual will tell you what's involved. B suffix engines have a slightly meatier crankshaft, and all that that implies.

While you've got the engine out, tidy up rough edges of machining and remove excrescences from within castings. They could chip off and cause damage.

It appears that the earliest (1958) roller-style cam followers for the inlet valves were finished in white metal with the rollers slightly tapered. It is commonly held that the large bore through the roller, the chamfer or the larger indentation, marks the narrow end of the taper and that it should face forward, but in practice only a micrometer will show it up for certain. Some aftermarket rollers, and maybe some factory originals, had a large indentation on one face and the smaller on the other which did not necessarily correspond with the taper. Be warned!

Solihull later changed the inlet valve cam followers for a revised pattern incorporating larger cups for the pushrods, and at some time long after the final P4 saloon had rolled off the production line the cam followers underwent further modification, the final pattern being finished in a green coating with cylindrical rollers, not tapered.

In 1959 BSF gave way to UNF for the tappet adjustment – it's certainly something to be aware of and look out for. UNF was an addition to the BSF and Whitworth ranges which continued throughout the P4's production life.

Check that the oil holes in the rocker shaft are located so as to correspond with the rockers and that they direct oil at the point where the maximum effort, and thus wear, occurs – and not directly at the valves. The drilling seems to vary between different examples. Also check that the inlet rocker shafts have threads for the locating screws. Check carefully for swarf before fitting.

Both phosphor bronze and steel gears were produced for the distributor and oil pump drive; the last range of Land-Rover 6-cylinder engines had steel gears. Check for end play which would, if not taken up with appropriate spaces, produce differential ignition timing between idling and under load and result in needless wear on the drive gear. The Workshop Manual warns against replacing steel driving gears with the phosphor bronze type unless a new camshaft is fitted; should you ignore this warning the resulting assembly will be short-lived.

Check the exhaust valves for substance and, if so worn as to have a sharp edge and little or no 'meat', replace them; they would reach extreme heat and cause pinking and suffer undue wear.

The original inlet valves had fairly short guides with a rubber O-ring oil seal located halfway down the guide: not the finest design feature of the engine. For the later 4-cylinder Land-Rover engine similar sized valves were used, but with longer guides and with the oil seals at the top of the guide and easy to access – altogether a better arrangement. Consider investing in modifying the cylinder head by fitting the later Land-Rover valves, guides and seals.

If you're dealing with an engine with that notorious O-ring between the thermostat housing and the water pump and you have to skim the cylinder head, beware of what effect your efforts may have on that O-ring!

Even if you take a brand-new factory original oil pump from its box, check it for sharp, unfinished internal surfaces which could release swarf and ruin either the pump itself or whatever other spot it might reach.

Don't assume that new piston rings will fit perfectly; it may well be necessary to grind some or all of them to produce the requisite gap – if it is not done the cylinder bores will become glazed and horrendous oil consumption will result. Moral – take nothing on trust, even if new and in the maker's box.

Check that the oil filter bypass valve ball bearing is not jammed part-open by swarf, thus allowing unfiltered oil to circulate.

Look out for any kinks and potential weaknesses in the oil feed pipe taking oil at high pressure externally to the pressure-reducing valve at the inlet point to the rocker shaft – and be aware that the drilled bolt which passes through the banjo can become distorted and prone to fracture if ever it has been over-tightened. Always fit the correct new copper washers on reassembly.

When replacing the main bearing caps, check that the old spring washers have not cut into the caps and the block and consider replacing the spring washers with suitable flat washers.

Consider fitting a plate above the oil pump to limit the amount of oil thrown upwards, and consider partially baffling the sump to ensure that on deceleration the bulk of the oil stays towards the rear where it will be picked up, strained and passed through the pump rather than being flung forward so as effectively to starve the pump.

The water pump housing is almost bound to be corroded around the edges. It can be subjected to a light skim, with an equal amount from the impeller surround within. Rotate the impeller on a lathe to make sure that the boss is drilled centrally, you may have a nasty surprise. On reassembly the gasket should have sealant applied around its internal perimeter to prevent it becoming soaked in coolant. The modern practice of maintaining 50% inhibited anti-freeze at all times should of itself substantially reduce proneness to further corrosion damage.

The P4 is already substantially over-provided with cooling capacity for the UK climate and has no need for a conventional fan which works hardest and is noisy when least needed, and is at its least efficient when most needed, why not fit a Kenlowe thermostatically controlled electric fan instead?

If there are cooling problems, attend to corrosion sludge build-up in the block and have the radiator professionally rebuilt: a surprisingly inexpensive operation. They should not recur for many a year provided 50% inhibited anti-freeze mixture is used all year round.

The timing chain housing as installed in the 3-Litre and Land Rover

Series III has been re-engineered to enable the front oil seal to be accessed from the front; this is a much better arrangement than that in the 95, 100 and 110.

The distributor from the Rover SD1 2600, which has both vacuum and centrifugal advance and employs points-free electronic timing by magnetic induction can be adapted to fit if the lower body is put through a lathe; once the timing is set it remains constant. Only the amplifier will need to be relocated and that will present no problems.

Some of these ideas may seem rather wild and extravagant and can be taken or left to choice, but most of them are plain common sense in the light of today's knowledge and will result in much longer life between rebuilds than ever had been in contemplation when these engines first hit the market 45 years ago, and when the cars were known as the poor man's Rolls-Royce. Millers Oils even suggest that if one of these engines is rebuilt with attention to every detail it can, after a proper period of running in, be run using an appropriate grade of modern semi-synthetic oil.

45 years ago, 100,000 miles between rebuilds would have been tolerable, even for a Rolls-Royce; nowadays 500,000 is not unknown in a BMW 7-Series using modern lubricants. That is what has inspired these suggestions and what they are aiming at. So please don't think I'm trying to suggest you make a racing Rover P4! I just wish you every good fortune in achieving today's life expectancy from what was considered a top-class engine all those years ago.

Brian Griffiths ***May 2004***

WINDSCREEN WIPERS

Whilst waiting for the distributor to be overhauled by Holden Vintage, I thought I would explore the innards of my Rover 95's windscreen wiper gear. The Workshop Manual is clear and helpful, and the task is fairly straightforward.

I first removed the cover plate of the motor unit. Inside was the original grease, virtually solid. I was sure that this could not result in easy movement of the parts, so I proceeded to strip them down. It is quite tricky to remove the plate between bonnet and windscreen. The passenger side wiper arm support does not need to be removed; the plate is probably adherent to sealant from the windscreen rubber; and it helps if the fresh air flap does not have Brewer's Droop. The flexible drive cable is continuous and runs through the metal pipes, smeared in stiff, old grease. When the pipes and drive units are dismantled, four studs remain, which are like miniature engine mounts. In common with engine mounts, the rubber seems to be prone to cracking. I tried to renovate mine by levering open the worst cracks and putting in 3M's black rubber adhesive. I wonder if new replacements are available? The drive units are easy to dismantle and clean. I sprayed all parts liberally with white lithium grease, which I hope is an acceptable replacement.

Unfortunately, my spindle rubber grommets are partially split where they have a flange. I don't think that replacements are available. If anyone is breaking a car with intact grommets, I think that they would be worth preserving. I sprayed them with rubber and nylon lubricant to facilitate the fitting of a metal plate through which the grommets and spindles protrude.

Richard Redmond ***May 2004***

REAR WINDOW LIFTING CHANNELS

The rear window mechanisms of the P4 are more prone to rust and disintegration than those of the front windows, due largely to lack of use.

The rear window winders on my 1951 Cyclops would wind down but the windows did not move, after some slight banging by hand they both disappeared down into the doors. On stripping out the windows and the lifting mechanisms I found that while the actual mechanisms and the glass side support channels were in quite good condition, the lifting channels in which the glass sits had totally rotted out and I was left with lengths of rotten lacework, which were totally irrecoverable.

(See Fig.1 for drawing of the original glass holder)

FIG I

ORIGINAL GLASS HOLDER AND MECHANISM

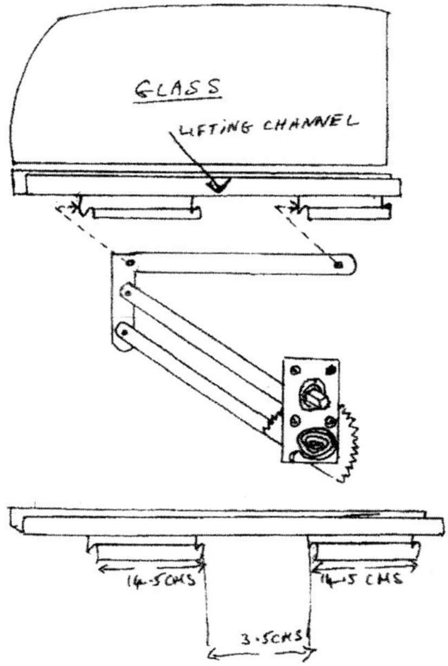

The problems was where to get replacement channels. I telephoned John Wearing, who has been marvellous in supplying the spares for the Cyclops. He told me that unfortunately the channels were not an item that was re-manufactured but that he would check some of his scrappage cars to see if he could find two acceptable channels.

Being located in Ireland and being impatient to remedy the problems, and particularly having made new door panels which turned out very well, I was anxious to complete the job on the doors with fully operational windows.

I went to my local hardware store and purchased a length of aluminium channel, which is used for securing the edge of parquet flooring. The channel, which is U-shaped, measures 30mm in height at the front, 20mm in height at the rear and 10mm in width. I was guessing that with a little ingenuity, I could make a new set of channels for the windows.

I first cut a piece of channel of approximately 520mm, (Fig II) the length of the rotten channel in which the window sits.

FIG II (REPLICA GLASS HOLDER)

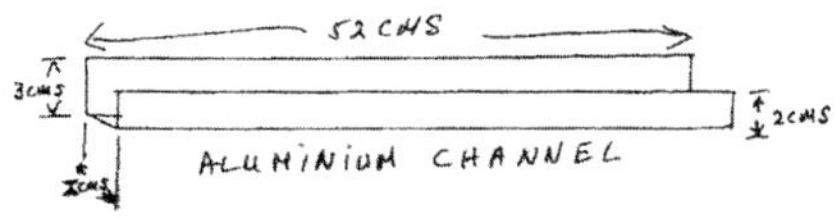

This would form the base of a new channel. I then cut two 360mm pieces of channel which measured approximately 40mm longer than the length of the runners in which the wheels of the winding mechanisms sit. (The excess metal permits riveting the two lengths together – see Fig. VI). The two 360mm lengths are required to form a box which, with approximate slots cut out, would form the runners in which to locate the wheels of the lifting mechanism.

FIG III WHEEL RUNNERS UPPER SECTION

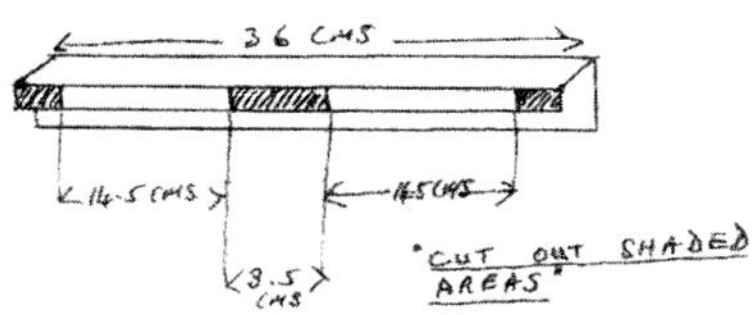

Taking one of the 360mm pieces I cut slots out of the 20mm side, which left projections equal in length to the runners on the original rotten window channel (145mm). Next the second 360mm piece of channel was slotted in a mirror image of the first one. (see Fig.IV)

FIG IV

WHEEL RUNNERS LOWER SECTION

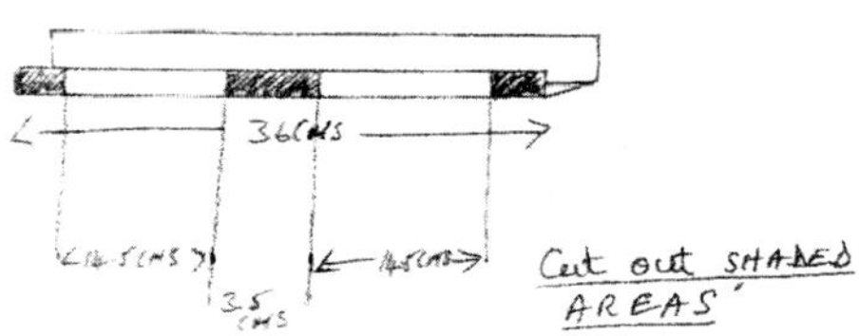

The next step was to pop rivet the base of one of the 360mm slotted lengths to the centre of the base of the 520mm window channel. (The slots permit the pop riveter to successfully get the rivets into the very narrow apertures in the aluminium channels.) (see Fig.V)

FIG V JOINING WHEEL RUNNER UPPER SECTION TO GLASS HOLDER

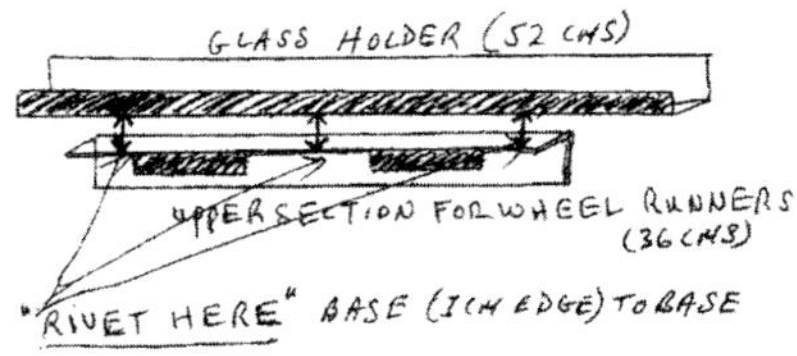

Step three involves pop riveting the second 360mm slotted length to the first 360mm length. The rivets are punched through the higher 30mm sides of each length, which are first overlapped. At this stage you now have a 520mm window channel with two slotted 360mm lengths riveted to it. (See Fig. VI)

FIG VI JOINING WHEEL RUNNER LOWER SECTION TO UPPER SECTION (AND GLASSHOLDER)

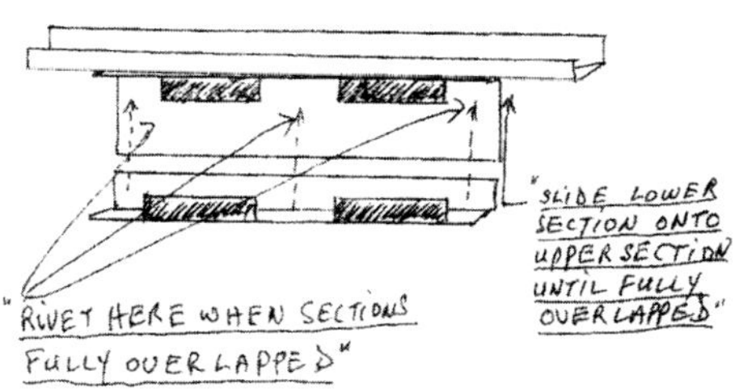

Using an angle grinder I cut out an opening in the two 145mm projections of the two 360mm slotted lengths wide enough to permit the wheels of the winding mechanism to sit and move freely left and right. This completed the construction of the replica window channels. (See Fig. VII). With the replica window channel and attached runners finished, using an appropriately cut length of old inner tube, I glued both the rubber and window glass into the holder using epoxy glue. Note it is much easier to glue the glass into the channel when the channel has been inserted into the door and the winding mechanism attached to it.

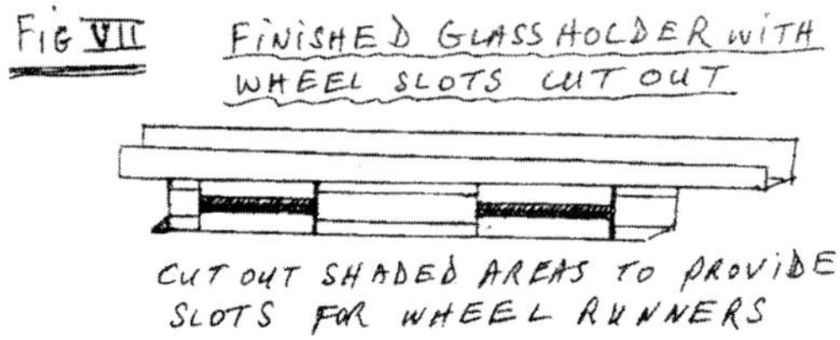

Then it was just a matter of greasing the new runners and the windows now work perfectly.

Aidan Geraghty ***March 2005***

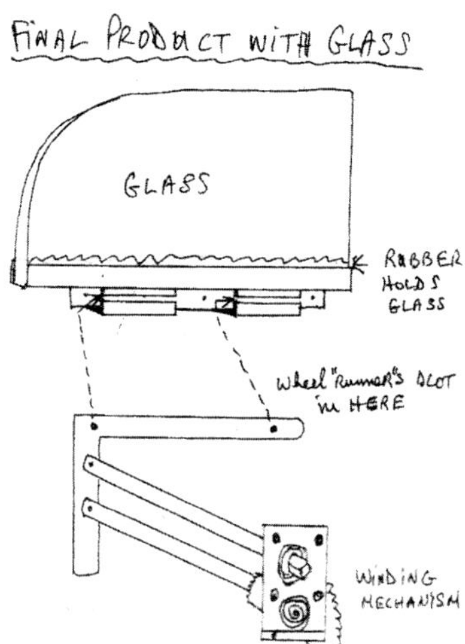

OVERHAULING BRAKE CALIPERS

In the official literature and instruction sheets issued with new seal kits, it says that under NO CIRCUMSTANCES must the calipers be split. I had a pair of old ones which had come off a car I had dismantled. Being an Engineer, and a nosey one at that, I decided to split one to find out why. On one side there is a recess cut in which to place an O-ring to seal the two halves of the transfer port from one cylinder to the other. The opposite side is plain, not recessed. A friend of mine was one of the draughtsmen in Girling's Drawing Office, so I asked him why it was so important not to split the two halves. His answer was: "It's a load of rubbish, provided that you have a new 0-ring to replace the old one!"

Only two days ago I had to replace the seals in my offside caliper so I cleaned up this old one, put a new 0-ring in the recess, gave one of the two surfaces a thin coat of clear varnish, keeping it away from the transfer port – so as not to block it - and pulled the four bolts up very tight!

After a run round the town, the brakes are excellent and the outside of the caliper is bone dry.

It is so much easier to fit the dust boot when you have got the two halves separated. The depth of the recess is 1/16" so you will need an O-Ring 3/32" thick to allow for compression. The diameter of the recess is 15/32". If you can't get an 0-Ring, and you have got a 7/16" wad punch, a piece of 3/32" thick rubber with a 3/16" hole through the middle will do just as well; this is what I used.

The four bolts holding the thing together are very tight, you will definitely need a piece of pipe on thc cnd of your socket tommy bar!

Tom Morgan ***September 2005***

(Tom tells me that the rubber in the middle of a wheel cylinder seal is the correct thickness; just remember to bin the remainder of the seal afterwards so nobody's tempted to use it – Ed)

THE HT CIRCUIT

There are two types of rotor arm most often found in the later six-cylinder P4s. The pear-shaped one left is not serviceable as explained below. Of the two, most of the experts seem to favour the alternative, fob-shaped model.

Using the pear-shaped rotor arm a pimple can occur which can cause it to spend its (short) life riding too high on the shaft, thus forcing the whole of the spring-mounted carbon brush (which is supposed to project) deep into its housing. If a pimple does appear this may cause a short circuit of the HT to earth, probably as a result of the arm not having been a snug fit on account of the pimple.

On both arms the length of the arc of the brass blade is long enough to enable the rotor arm to cater for advance and retard, without being so long as to affect the firing of the next plug in the sequence.

At a recent meeting of the Cheshire & North Wales group we looked at several distributors and saw a variety of interesting sights: one car was running beautifully without there being a carbon brush and spring in the distributor - it must have had one hell of a spark! - but the owner serendipitously had it on his list for checking the very next day.

With the demise of Lucas it's extremely hard to come by definite information on original distributors and rotor arms and how they were selected; even the nomenclature varies... 25D, DMBZ 6A, 45D and so on. However the distributor used in our cars is by no means confined to Rovers: several Jaguars and Aston Martins to name but two other makers used it. From the line drawing in the Rover Parts Manual it would appear that the pear-shaped rotor arm was originally used for Rovers, however a fob-shaped rotor arm seems frequently to feature under members' car bonnets. Rover 110s and late 95s had a squatter distributor using a cap with side-outlets for the HT leads, whilst 100s and the earlier 95s had "high" distributors using caps with castellated tops and acorn fittings. However I understand that the rotor arms used are interchangeable.

It's well known that a large number of all the cars concerned, not just Rovers, have recently been suffering sudden catastrophic breakdowns due to HT problems. These are often blamed, rightly, on poor quality after-market parts, but as so many classic cars do very low annual mileages the seriousness of these problems may not have become apparent nearly as fast as might have been expected.

The quality issues seem to affect pear-shaped rotor arms, some of which seem to be made from a plastic which does not have the requisite electrical insulation qualities and some have a poor internal finish which causes them to ride high – by checking and if necessary rectifying before fitting the part the latter problem can perhaps be eliminated. Inadequately riveted brass blades which come adrift are far from unknown: I very recently heard of a Jaguar engine suffering very expensive internal mechanical damage as a result of wrong firing caused by a dislodged blade. The variety of rotor arms on show at the NEC Classic Car Show in November and the standard to which they had been constructed was extraordinary. Some had a locating spring immediately beneath the rivet - surely an invitation to find the easiest way to track through the moulding? Some rotor arms even had the locating lug moulded several degrees out of line. Generally it's a good plan to seek originals and to avoid after-market items altogether - the words cheap and nasty all too often come to mind. Jaguar/Daimler Heritage supply ignition parts for the 1958-61 XK150, the 1960-63 3.8 MkII and the 1961-63 E-type 4.2 which are boxed and guaranteed by them: wherever applicable it's probably a very good plan to use them whatever they cost.

Before fitting a proposed new rotor arm you may wish to place it on a flat surface alongside the original and compare their heights, since some aftermarket products are incorrectly dimensioned and stand too high on the shaft. This is quite apart from any pimples or internal blemishes which have not been rectified prior to despatch from the factory.

When fitting a new rotor arm it's a good plan to check its dimensions against the old one, and once installed to ease the distributor cap into position ensuring that you can feel the brush retaining spring being gently compressed, then checking that the act of compression hasn't actually pushed the brush into the housing beyond recall; also watch the distributor cap for movement, listen for noise and feel for heat as some distributor caps and rotor arms don't marry up as well as they should due to differences in manufacturing tolerances.

Holdens have recently started to remanufacture side-entry distributor caps to the earlier 1950s pattern and the associated rotor arms, and they tell me they intend to address our needs soon with components which they assure me will be made to the proper specification; I hope this will prove to be the case and that the rotor arms will be of the fob-shaped variety. In any case members will have to brace themselves to meet the realistic cost of manufacturing for, and supplying to, a relatively limited market. I venture to suggest that the Anti-Classic-Car lobby would love to cash in on the danger which breakdowns can cause etc. etc., blaming old cars and not poor quality modern parts, and that we ought to grit our teeth and do our best to make sure our cars don't offend even if it's not cheap.

In the interests of reliability and minimal maintenance I have fitted Pertronix Igniter (which is identical to Aldon) and am delighted with it, as are all those I know who have done likewise whatever their classic car. I've also fitted it to the Sl Bentley which has a Delco double-points distributor - what a joy not to have to adjust them again! The Lumenition system is also well spoken of. You need to specify whether your car is Positive or Negative Earth and if you mention you're a Guild member you may be able to get a discount. Beware of the Lucas Opus system, however, as it soon earned the nickname hopeless! These systems are "points-replacement" and not "electronic ignition" hence the original (3 Ohm) coil should continue to be used.

Carbon fibre HT leads reduce interference on FM, and even if your radio doesn't offer FM the person next to you at the traffic lights may well be cursing your "old banger" if you're not using them. With 'acorn' connectors in the distributor cap and conventional carbon fibre leads the problem is to get a physically robust connection. Pertronix and Aldon supply plain black 7mm leads which have a centre core of carbon fibre contained within a twisted steel core: thus they are flexible, suppressed and you can make a secure connection onto the copper washers within an acorn cap: having fitted them I'm very happy with them.

If you're using a Sports coil why not consider if you really need one? With the P4's relatively low maximum revs and famously good starting it really shouldn't need a Sports coil, and by using one you are subjecting the HT system to higher voltages and thus greater likelihood of breakdown. Mine has gone into the boot for spare.

The traditional plug caps look very nice and are well worth keeping for show purposes but they can be a significant source of 'tracking' which leads to misfiring, and to use modem plug caps for everyday use is certainly not a crime.

And in closing it's worth noting that people have been reporting many problems with recently purchased after-market coils, even those in Lucas boxes. Consider paying the price for a specialist product – again Pertronix or Aldon spring to mind amongst others.

Brian Griffiths ***January 2006***

THE PETROL RESERVE PICK UP

The petrol reserve on my 100 ceased working and in spite of stripping the solenoid and repairing a disconnected wire no improvement was forthcoming. Having two fuel pumps, the obvious solution was to use a two pipe system from the tank in place of the solenoid and single pipe and therefore after a natter with Stan, I obtained a 95/110 pick up system from Richard Hedger. Although there was petrol in the tank, the new system would not pick it up and I realised the new pipe was not reaching the bottom of the tank.

To get accurate positioning of the pick up, I made a wooden jig replicating part of the tank and found that the reserve pipe was some three quarters of an inch from the bottom. Bending the pipe to suit cured the problem.

The later pick up consists of a half inch copper pipe with a smaller diameter plastic pipe running through it and protruding about an inch from the bottom. Plastic fuel pipes were used to feed the pumps with two elbows on the tank unit and the two existing elbows reused at the pump entries. I could have used straight connections at the tank but due to the restricted space around the pumps, elbows are necessary here. An additional hole is required in the boot corner for the extra feed pipe.

Derek Humphreys ***July 2006***

BRAKE FLUID – SYNTHETIC OR SILICONE?

The question I am often asked is: 'what brake fluid should I use in my P4?' – ie standard mineral fluid, synthetic or silicone. The answer is, you can use any of the three, but whilst standard mineral fluid and synthetic are compatible – SILICONE FLUID MUST NEVER BE MIXED WITH ANY OTHER TYPE OF FLUID! In fact the only way that silicone brake fluid can be introduced into a braking system is when the entire system has been built from scratch with every part – slave cylinders, calipers, master cylinder, servo, flexibles and all piping being brand new! It's simply down to chemical compatibility and silicone mixed in any way with conventional brake fluid, in any proportion, will result in all the rubber components becoming swollen and eventually being unable to slide in their component parts.

The DOT (Department of Transportation) numbers 3, 4, and 5.1 are all suitable for the P4's braking system, but be sure that you know which fluid has been used in your vehicle. Depending on how much you pay for the fluid reflects in its service life. That is to say, the lower priced item will be safely usable for about three years, ranging from five to even ten years for the much more expensive fluids. Personally, I think brake systems should be completely replaced with new every two or three years, if only to 'exercise' the bleed nipples. The reason I say this, is the fact that standard fluid is hygroscopic and therefore has the capacity to absorb moisture into it and eventually corrode the inner workings of the brake components. Silicone fluids are non-hygroscopic but the very action of the brake system particularly in hard driving can actually create moisture by the fluid boiling and thus making steam, probably the only downside of silicone brake fluid.

Non-silicone synthetic brake fluid is fully compatible and miscible with conventional brake fluid. Never re-use brake fluid because the old adage of allowing used fluid to stand overnight to allow the air globules to dissipate is not true and in fact, while this is supposed to be happening, the fluid will have soaked up more moisture. Always buy fresh fluid when carrying out brake work to your car as fluid standing on your garage shelf could have been there for longer than you think!

To go back to my point that silicone fluid mixed with conventional will cause the component rubber parts to swell (like a human bodybuilder using steroids!) I must relate to you the story of one of our members decrying one of our parts suppliers by saying he had sent him a faulty brake master cylinder. He asked if he could send it to me for appraisal. On receipt of the obviously brand new cylinder i.e. not 'old new stock', I proceeded to strip it down for inspection, on removal of the actuating rod and the large end nut I discovered I couldn't remove the piston. With the use of a brass drift, I finally managed, dare I say it, to hammer the piston from the bore of the cylinder only to find the end seal had swollen to almost double its normal size after being extricated from the bore. You could almost hear it sigh with relief on its release!

This was simply due to a chemical reaction to the two very different types of fluid being introduccd to each other. Vendor vindicated, customer totally gutted, when I told him the problem. To be fair to him, it was not his fault. He shyly admitted that his local garage had carried out the work. There's a moral there somewhere, I think.

So be warned, the introduction of silicone fluid to glycol and polyglycol (synthetic) to the system will start off a chemical war of its own and apart from swelling of rubber, as described, the silicone will generate a sludge that will seriously affect the recuperating holes in the master cylinder and the very small seals in the brake servo unit (if one is fitted). Once silicone has been introduced to a system that has been happily using conventional fluids, there is no going back. Just bleeding and refilling the system with 'ordinary' fluid will not suffice in any instance, so unfortunately the entire system will have to be stripped and every single rubber part will have to be replaced with new. This includes the three flexible brake hoses unless you are lucky enough to have stainless steel braided hoses with a non-rubber sleeve made of Teflon.

Stan Johnstone ***May 2006***

ROVER HEATER OVERHAUL

Ever since I've owned my 110 to say that I have been disappointed with the output of the heater would be an understatement. It was on a run from Pinner to Maldon for a Spares Day in October 2004 that I decided that 60 plus miles to get any heat to the passengers was simply not good enough.

A careful examination of cooling/heating system when thoroughly warm revealed two things. Firstly, there seemed to be a relatively cool spot in the middle of the radiator, which probably accounted for its tendency to overheat when stuck in Motorway jams (I had put this down to the fact that when the P4 was designed traffic jams of today's magnitude were never contemplated, although on reflection as P4s were exported to hot countries like Australia this was a false assumption) and secondly, whilst the feed to the heater matrix was hot, the pipe from the matrix was only warm.

This seemed to indicated some sort of blockage in both the radiator and the heater matrix. At this stage I must admit that domestic pressures put the matter on the back burner and any further investigation was delayed until some months later when a leak from the header tank forced me into taking action. I removed the radiator and took it to a local specialist firm in Harrow who traded under the name of Viking Radiators – good name. I was further encouraged when the man behind the counter took one look at the radiator and said: ‘Ah, looks like a Rover P4 rad. Got a leak at the header tank? Always was a weak spot.’ He undertook to have it fixed by the next day, however because of the cool spot I asked him to check the flow rate and sure enough this turned out to be substandard. Whilst he managed to shift a lot of lime scale he could not get a satisfactory flow rate and reckoned that an area in the centre about the size of a man’s hand was completely blocked. The only certain way to resolve the problem was to recore the radiator, which took about a week and cost just over £100. It was more than I had budgeted for but if it solved the overheating tendency then it was money well spent.

This of course did not solve the heater problems although the temperature gauge now stayed resolutely just on the lower side of normal and overheating seemed a thing of the past, the performance of the heater remained pathetic. It was pretty obvious from earlier investigations that the heater matrix was also blocked. The removal of the matrix is fairly straightforward and it turned out to be completely blocked. A weakened solution of household limescale remover was poured in one end and left overnight but had little effect, so I bit the bullet and tracked down a matrix from one of the usual Rover spares men which he was able to supply exchange at a reasonable cost. In fact it turned out to be a new ‘old stock’ item. The swap was relatively straightforward and I looked forward to warm feet. A short road test proved how wrong it was to make assumptions. Whilst hot water was definitely passing through the heater somehow little or no heat, even with the aid of the blower, was getting to my feet. Logic had it that the fault must lay between the heater matrix and the heating ducts so there was nothing for it but to take off the blower. This was a simple job, just four screws from inside the car, loosen the clip holding the rubber sleeve, which joins the blower to the scuttle, and the whole assembly came off. The blower can be stripped quite easily and it was immediately obvious that here was the cause of the problem. At the back of the blower is a circular piece of wire mesh, no doubt to stop debris from getting into the heated matrix. This mesh was completely blocked with dirt, in fact at first glance it looked like a blanking plate. In addition each of the blades of the blower was caked solid with dust, in some places up to almost a quarter of an inch thick. All in all a bit of a mess. Strangely the electric motor itself seemed ok, perhaps centrifugal force and the sucking effect of the fan blades had kept most of the dirt out. It was not possible to strip the motor as the two halves of the case are riveted together, however after a quick trip to a local garage which kindly let me use their airline to blow out any residual dirt and a bit of judicious lubrication, all seemed fine. I also used the flexible tube from a Hoover to suck out any dirt from the ducts and as an additional measure used the airline to thoroughly blow through the system. The wire mesh was soaked in a solvent and brushed clean as were the blower blades, taking care not to dislodge the small balance weight. Reassembly was straightforward, the whole job taking no more than a couple of hours from start to finish. I now have heat. Not to the standard of a modern car but enough to raise the temp from 50 to 70 degrees F in about 15 mins when the external temperature was hovering around freezing.

On reflection I would hazard a guess that the blower had not been cleaned since the old girl came off the production line. Anyway I can now look forward to winter runs without my thermal socks.

Malcolm Lowe ***September 2006***

SEAT BELTS IN A P4

At the recent NEC Classic Motor Show, we faced one question more often than any other: 'Can you fit seat belts to a P4, and if so then how?'

Maybe the latest regulations for children and belts have focused attention on the subject – for people of all ages. On the subject of children in cars, the new regulations are explained at http://www.childcarseats.org.uk

Of course most P4s never had belts, and in a bump between a 1.5 ton P4 and a lighter car you're less likely to be shaken around in the P4 – think of a collision between a snooker ball and a table tennis ball to understand why. Other cars on the road are getting heavier all the time though, and many studies have shown that on the whole, the risk of being injured in an accident is roughly halved if the occupants are wearing belts. This risk is especially reduced in frontal accidents – the most common type – and most of all if a car rolls onto its roof.

If your P4 is a pristine, original show-winner then you may prefer not to alter its looks by fitting belts. After all, you don't legally need to have them unless your P4 is one of the few 1965-registered models. I bought my P4 to use rather than to win prizes though, and before putting it on the road – and certainly before my sons travelled in it – I looked at the options for fitting belts. After living near a major road junction before compulsory belt wearing from 1983, I needed no persuasion.

The Front

Unusually for their era, later models of the P4 left the factory with mounting points for front belts. Inside each B pillar between front and rear doors, behind the vinyl trim and its hardboard backing, you'll find two holes at shoulder height. Rover supplied a thick steel bracket to run between these two holes and carry the shoulder mounting by means of a hole in the middle of the plate. If your P4 doesn't have the bracket, it shouldn't be difficult to get it second hand.

I noticed these holes soon after buying my P4 as an interior-stripped project. Even better, the boxes of interior parts contained a pair of belts. I was disappointed, though, to realise that these were only the two point diagonal type; fine in a bump for keeping my rib cage away from the hard steering wheel boss, but they wouldn't prevent my lower body hitting the area under the dash. I also wanted inertia reel, rather than static, belts. Purists may scorn – inertia belts didn't appear until the mid-1960s – but I remember the difficulty in reaching the minor controls of my first car – a 1967 Vauxhall Viva HB – if I wore its static belts.

My next effort to find suitable belts didn't work, either. At a car show a couple of months later, I thought I'd picked up a bargain in the form of two pristine, new old stock Ford belts. However these were simply too short for the P4's tall interior. There was only one solution. Get some tailor-made belts. Fearing a high cost I contacted Quickfit Safety Belts, after seeing them mentioned in Practical Classics. For a very fair price they made a pair of front belts to the length I requested. I'd worked out the length needed after measuring between the shoulder mounting, the central mounting near the transmission tunnel and the new mounting needed near the sill. The latter can either be simply bolted to the floor by a bracket, as in my car, or alternatively it could be attached to the bottom of the B pillar. Just keep your eyes open for rust in this area! If your P4 has two-point belts, you'll need another mounting hole for three-point belts.

Besides supplying the belts, Quickfit (not to be confused with the similar-sounding national tyre, exhaust etc supplier) also sold me enough nuts, bolts, washers, cushioned bolt covers, brackets, spacers and thick steel spreader plates. The latter part is essential, even with the P4's thick bodywork.

The complete kit of parts proved straightforward to fit, and the belts are comfortable in use. My only reservation is that the P4's shoulder mounting is a little low, causing the belt at times to slip off my shoulder and down the arm. Perhaps Solihull designed it for two-point belts across the body's centre of gravity. Or maybe I should put the seat bolts through different mounting holes to lower the seat, and lose two inches in height to be on a par with the average 1957 driver! Either way, I'm tempted to buy a new pair of steel brackets from Quickfit, as they now make replicas which place the top mounting at the correct position of one or two inches above the top of the occupant's ear.

Quickfit can supply the floor-mounted part of the belt as either a flexible stalk or a webbing strap. For a small extra charge they can also offer a choice of webbing colours, to match your P4's hide.

You may be lucky enough to find a set of original, period front belts; these will look 'right' for the car. They're thin on the ground though, and I'm wary of second hand belts anyway. Many people don't realise that belts are designed to stretch in an accident and cushion the impact. Once stretched though, they stay stretched and won't give any cushioning a second time. In an accident the P4's rigid chassis shouldn't distort as much as a modern car's crumple zones, but the trade-off is a greater rate of sudden deceleration. This is often the major cause of life-threatening injuries to internal organs. All the more reason for the cushioning of new belts with some 'give'. Can you be certain that those 'bargain' used belts haven't already stretched? Also if they are at all frayed, they'll fail an MoT. If you insist on period-looking belts, Quickfit can supply new static belts with old-style webbing and buckles – although these aren't recommended for use with a child's booster seat. They have produced a novel solution to this problem, by offering to make a belt with both a period-looking buckle and a modern tongue.

The Rear

It's easy to dismiss rear belts as being less important than those in the front. After all, won't the front seat backrests cushion rear passengers in a bump? They probably won't: look at the twisted backrest of any front seat after a rear passenger's been thrown at it in even a modest bump. Then bear in mind that front and rear occupants' heads can also collide.

The good news is that rear belts can be fitted in a P4. It's more of a challenge than in the front, although I've seen several approaches amongst Guild members' cars.

1. Rover's approach

Even more remarkable for a 1950s design, Solihull actually designed a rear belt layout for the P4. It allowed the upper mountings of static belts to be fixed to the boot hinge mounting. Details were given in the March 1998 edition of Overdrive. I'm assuming that by now the special bracket to attach the belt to the boot floor is extinct, although the upper mounting can be simply bolted to the boot floor or inner wing so long as a thick steel spreader plate is employed. The belt then passes between the backrest and parcel shelf, whilst the lower section of belt is simply bolted to the floor, near the point where the backrest and squab meet.

2. Exposed mounting static method

At least one member's car uses thick steel supports lengthways under the parcel shelf to carry the upper mountings of static belts. The shelf itself is of no strength whatsoever. Each steel support can be a few inches from the edge of the shelf. This method avoids disturbing the boot area with belts – not that they take up much space if they pass immediately behind the backrest – although the downside is a need for welding near the rear window. Ideally you should probably remove it first. No thanks unless you want to create a large amount of work.

Both these methods can be used to provide three-point belts to the nearside, offside and the middle. How many modern cars until three or four years ago boasted a three-point middle belt in the rear?

3. Quickfit's inertia reel method

If you want inertia belts in the rear, Quickfit has devised a method. This mounts a reel discreetly to the inner wing inside the boot, from where the belt passes though a neat hole in the rear shelf. If you wish to adopt this approach on a DIY basis, the biggest challenge is to pass the belt through the rear shelf in a manner which is strong enough yet prevents the belt from getting scratched and fraying. Not an easy method, but neat and effective if you've got the skill to make it. Again, the bottom mountings follow 'Rover's method'.

4. The author's method.

I wanted inertia belts in the rear, but wasn't convinced I had the skill to adapt Rover's method for inertia belts. Therefore I used the existing, thick supports which run under the rear shelf – about six inches either side of the centre of the car – to carry inertia reels on the shelf above. These supports looked strong enough to support belts for children and probably also adults, although this mounting position was never tested by the factory. To be certain of their strength I reinforced the point where the rear of the support joins the bodyshell, using a bead of weld to supplement the two thin bolts. The belts are a standard, horizontal mounting type as supplied by Quickfit for open sports cars such as the MGB.

It's essential to use spreader plates under the shelf supports, and to fit spacers between the support and the shelf itself.

The result is comfortable for children of all ages. The belt should really be adjusted to run across the tops of the legs, not across the lower abdomen. This is probably the easiest way of fitting inertia reel belts.

On the downside, I think it looks ungainly; the narrow rear shelf of the P4 doesn't absorb reels as well as, say, the wider shelf of a P6 on which this arrangement was sometimes used. Incidentally, it's legal to mount the reels on the inside of the car and the buckles on the outside of the shelf; BMW used this layout on their early 1990s 3 and 5 Series.

I don't recommend putting the reels above the shelf. This would leave the reels too close to rear passengers' heads, and could cause injury in a rear shunt.

5. Lap belts

A two point lap belt can easily be fitted, by drilling mounting points into the floor, beneath the point where the squab and backrest meet. I used this method for a centre belt. This is a simple, cheap and effective system for children and short adults. However I don't recommend it for tall adults, as I'm not convinced it would prevent them hitting their heads on the front seat backrest in a high-speed bump.

Lap belts of this type can easily be tucked out of sight under the cushion when not in use, and are ideal if you want to keep the car's original looks.

If you'd rather not fit belts yourself, Quickfit can make and supply belts to your specification for fitting by a local garage. Alternatively they can arrange fitting at extra cost at their Middlesex premises by prior appointment.

Tim Pearson — ***January 2007***

FUEL TANK FILLER AND ELBOW

Whilst working under the rear of my 1955 90, I noticed the fuel tank venting hose was delaminating, so I made a temporary repair with tape. However, it worried me, so I recently decided to replace it and the rubber elbow on the fuel filler neck.

The job is straightforward, if a little awkward, but I did encounter a few other problems, as follows. The spare wheel tray needs to be removed to provide access and if the clips are old, they will need cutting off with a junior hacksaw. It is also best to remove the filler neck and cap assembly, which is simple and obvious, with fixing screws inside the boot and extraction through the petrol flap aperture. You do not need to drain the tank, but be careful and take at least the following precautions. Once each hose is removed from the tank, keep the tank openings closed with thick plastic tape or blanking caps until you are ready to fit the new hoses. Remember, even an empty tank is a potential vapour bomb. Also, try not to let any 'muck' drop into the tank and never do anything to create a spark, when the tank openings are exposed. Keep the garage well ventilated and preferably do the job outside. The first snag is getting the correct size fuel hose for the vent pipe. It is 7/16ths or 11mm, but my local motor factor could only supply 10mm, which is too tight, but will have to be used. I managed to open the bore out with an 11mm grinding wheel in my "Dremel" high speed 'multi tool' and I also smeared a small amount of silicone grease on the pipe ends to help the rubber hose slide on. Water and/or Fairy Liquid would probably do. The hose is fairly thick, to cope with modern high pressure fuel systems and you may have to ease out the small hole in the boot with a file, really to just remove the rusty muck and underseal from the opening.

I already had a new rubber elbow which fitted over the filler pipe. There is a swaged ring formed at the end of this pipe and I did get the hose on enough for the hose clip to clamp the hose beyond this ring or collar. Again, I used silicone grease to help the hose on. The hose extends well up the pipe – almost to the boot floor and the flow of fuel down the pipe runs into the hose; it therefore will not leak out of the joint.

Fit and secure the elbow onto the tank pipe first and then insert the filler neck into the elbow, through the boot floor and secure it with another new Jubilee clip. Make sure all the clips are tight, and replace all the fixings for the pipe and spare wheel tray. Check for leaks, especially after first filling the tank, but I'm sure all will be Okay.

After I had removed the rubber hose and elbow, I discovered that the hose has completely collapsed. The elbow was even worse. Where it was secured to the pipe, it was still in good condition, but where the inside was exposed to fuel and fuel vapour, it had all broken up. I suspect some bits may have gone into the tank and I did notice grey/black silt in the fuel filter and carb when I recently overhauled them. It has to be assumed that unleaded fuel (coupled with age), is responsible for this breakdown of the rubber.

The job should have taken a couple of hours, but it took a couple of (part) days, because of all the fiddling about with the hose alterations. If you have to remove your tank, fit new hoses as a matter of course, plus the job would be easier to do. However, unless you know the hoses have been renewed in the last 10 years, or so, I would suggest early inspection and probable replacement, is a good idea.

Colin Gray ***May 2007***

RECTIFYING A POOR HANDBRAKE

Recently, I decided I was no longer going to tolerate the sloppy performance of Connie's (i.e. my 95's) handbrake. I had become weary of having to grit my teeth when parking on an incline, and making sure the wheels are pointed in a manner so as to keep the car out of trouble if it jumped out of reverse gear! If this is a familiar feeling, you may like to read about what happened when I decided to tackle the issue.

I have had P4s previously whose handbrakes have been poor, and I must confess to becoming a bit blasé about it – and thus I had put off this performance improvement project for a while. The thing is, Rover would not have designed a system that was ineffective, and with this in mind I decided to have a really good go at it so I could get as close as possible to the performance that the system would have displayed when new. Of course, I had previously done the stuff we can all easily do, i.e. adjusting the crook to two clicks only, and ensuring that the rear wheel cylinders could slide on their backplates. So when I looked further it was not surprising to find that the cross rods running along the back axle were bent in several places, the link plate was twisted and the relative geometry of the whole rear assembly was hopelessly out of true. Luckily I had access to a decent assembly on a spares car I had bought recently (what a 'domestic' that purchase caused!), and I was able to strip this off the car, disassemble and paint in a leisurely manner. It did take two weekends but as I say, I was in no rush and it was nice to thoroughly clean the bits and make sure all nooks and crannies got a decent coat of Chassis Black paint.

Upon reassembly, with all the bits coated in coppaslip, I was looking forward to getting it onto old Connie's rear end, and reaping the benefits. I was hoping to get it all done in a day and indeed, I thought I had done – it took a couple of hours to jack the car up, remove the wheels and get the old assembly off, with fiddly split pins taking the most time. (NB – I used a 1.5mm drill bit to drill out some of these, as some of them disintegrated during the removal process). Here's a picture of the old assembly at the side of the refurbished one – note the twisted link plate and distorted rods – evidence of brute force on top of poor servicing, I expect. No wonder it didn't work!

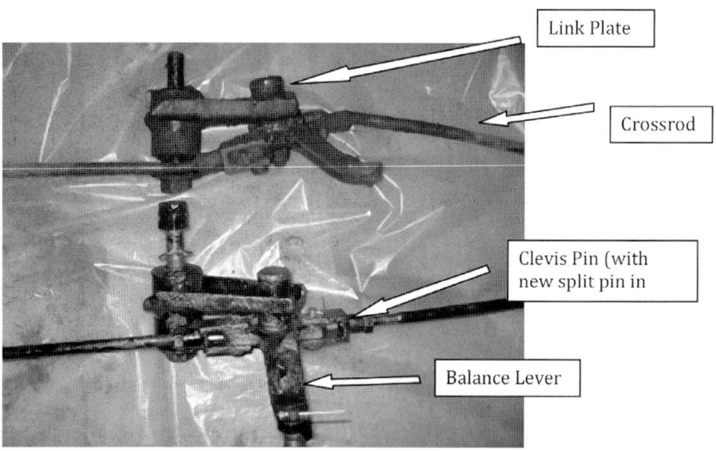

The refitting onto the car went well, although I had to finish off in the rain and the dark – winter is not the best time to do this kind of thing! Hence I was very fed up off when I checked the handbrake only to find it was worse than the previous set up. A good night's sleep was called for at this point, after which I was up for more investigation, adjustment and ultimately achieving the efficient system I knew was in there somewhere.

Whilst trying to reason out why the system wasn't working right, I realised that the geometry of the link plate and balance lever must be quite critical, and if you study the workshop manual you can see that the angles are cleverly specified to give plenty of travel for the crossrods. This allows them to pull the drawlinks (in the rear brake cylinders) totally out and thus fully force up the tappets that actuate the brake shoes, giving you a powerfully gripping handbrake. So, the next day, armed with my daughter's plastic protractor(!) I ventured under the car to set up the critical angles. Although this allows only approximate settings to be gauged, it's good enough because ultimately the accuracy of the angles you can set is governed by the coarseness of the crossrods' adjusting mechanism, ie the clevis pins which swivel on the threaded ends of each rod in the mechanism.

Being honest, I had hoped that all the angles would magically be right upon refitting, because I was secretly dreading having to adjust the length of the crossrods. However, I realised when I got to this bit of the job that almost all the relative geometries are actually only set and held true by the relative lengths of all the rods in the handbrake assembly - so you cannot avoid adjusting the rods if you want to get all the angles right. Unless you are very lucky, that is. As with most things, once you get going it's not too bad and I soon had all the rods adjusted to give all the correct geometries. I found I had to go back as far as the idler lever (i.e. the part of the linkage that's housed within the middle outrigger) to get all the rod lengths perfect. Sounds like a pain, but with enough time and enough cups of tea, it's no real bother (rather fun in the end, actually) and you get a warm feeling knowing you've done the job right.

Upon finishing, my work was rewarded when finally I pulled on the handbrake to find that both rear drums locked up tight. After a test run and a little adjustment of the shepherd's crook, I am pleased to be able to say that Connie now has the best handbrakc I've experienced, on any car. Yes, the whole thing took two full weekends – but hopefully now it's all set up all I'll have to do is keep the brake shoes adjusted up and lubricate the linkage, to always have a superb handbrake. Also, personally speaking, it was great to work on such a nicely engineered mechanism. This is definitely one of the bonuses you get from maintaining your Rover P4 – of course, my missus and kids tell me that makes me an old anorak, but I really don't care!

If you need/want to do this for your P4, I'd recommend getting in a supply of split pins, a 1.5mm drill bit, a wire brush, some penetrating fluid, Coppaslip, Waxoyl (to protect the crossrod threads, once cleaned), and a workshop manual. An ample selection of spanners and screwdrivers will not go amiss either – the clevis pins take a 1/8th Whitworth (3/16th BSF) spanner which generally makes undoing them easier than trying to use a screwdriver in the slotted end of the pin. Put plenty of time aside too, so you don't get stressed because you end up having to rush to finish. Good luck if you have a go – hopefully you'll realise the same enormous efficiency gain that I did, and thus you'll be a happier and safer P4 driver into the bargain!

Russ Sowden ***May 2007***

HOW TO SPOT A WORN PROP SHAFT

You may recall that last autumn I had the misfortune to be trailered home from the Lake District Rally with a transmission fault – see November 2006 edition for details. The car involved was a P5 three litre automatic, but the following problem could equally apply to the P4.

Fearing the worst but hoping for an easy fix, I delayed addressing the problem until late February when the weather had improved. I had previously checked the prop shaft for loose universal joints and was happy that there was no movement in any. Joints, which have lost some or all of the needle rollers, are easy to detect by the excess movement at the joint and a pronounced knock as the car is driven, therefore I was doubtful as to whether the fault would actually be on the prop shaft. However I removed the prop-shaft complete, it being similar to the P4 incorporating a centre bearing. The centre bearing spun easily (they do sometimes seize) but the universal joint at the rear axle end had locked solid and despite applying pressure I was unable to move it.

The joint had been running dry for several years and eventually the needle rollers, which are contained in a bearing cage, broke up and locked the joint. Unable to turn on its journals, the U/J caused the prop-shaft to vibrate wildly, the vibration being transferred to the rubber mounted centre-bearing fixture. This, in turn, caused the car to vibrate so much that the doors were being severely shaken along with the occupants. The whole episode was quite alarming, with horrendous noises coming from under the car.

Early U/Js were fitted with grease nipples and regular greasing ensured that the joints were good for extremely long periods. Later U/Js, although drilled for grease nipples, had a small screw inserted into the thread where the nipple was originally fitted. These were not meant to be greased, but the screw could be removed and a nipple fitted into the left hand thread if you were really keen. The current type of U/J has no drilling for grease and therefore relies on the seals at the bearing cage ends to keep the initial filling of grease in, and water out.

The grooved bearing face when dry, can cause rusty needle rollers to dig in before breaking up, the opposite bearing face had also run dry and was in the first stages of failure. The two opposing bearing faces were both as they should be.

The moral of the story is either to remove or disconnect the prop shaft in order to test the U/Js fully, it may save an embarrassing moment on the road.

Colin Blowers ***July 2007***

BRAKE SERVO FAULTS – THE OPTIONS

A sticking brake servo. Most long-tem P4 owners experience this frustrating fault at some time – some owners have even needed to replace their servo more than once within a short time.

Chat to any group of P4 owners, take just a brief visit to the Guild's on-line Forum, or read a few back copies of Overdrive, and you'll realise just how common this problem is.

I've decided to look at the current options open to an owner in the event of a servo either not giving as much braking help as it should or, more often, causing the brakes to stick on.

You can still buy overhaul kits, and at between £30 to £55 depending on type (all prices are net of VAT, p&p); they can make sense if funds are tight and you're confident of your DIY ability. They don't require any special tools to fit, nor should the task of removing the servo, rebuilding and fitting it take more than an evening – although it takes a little longer if twin carbs are fitted, as these need to be removed to give access. You'll then need to purge air from the hydraulics (why do people call it 'bleeding'?)

Stan Johnson rebuilt the servo on his 95 a full twenty years ago and it's worked ever since. Amongst other owners though, I've heard varied tales of the success rate of DIY rebuilds. With this in mind, I decided to approach Jon Wadhams of JRW, and also John Wearing for their views on the alternatives. Although JRW sell overhaul kits, if an owner prefers to replace a servo they recommend using a quality reconditioned unit of the original design, apart from a modified seal. This costs £175 (exchange) for most types. Jon Wadhams writes: 'Before we introduced our new reconditioned servo service a few years ago, the success rate was not satisfactory, however, now we have sold well over a hundred units with no valid malfunction due to the reconditioning (we did have problems with one customer only which turned out to be his master cylinder).

Due to the good reliability now, I would not consider a new modern replacement for the following reasons:

1. They are often not of the same quality as genuine Lockheed/Girling units as many 'new' unit are being imported.
2. They spoil the orginality and look.
3. They require more time to fit with the fabrication of brackets.
4. They rarely offer the same boost ratio as the 2A
5. They rarely offer the same displacement volume of brake fluid as the original 2A

You will find that most sellers cannot give definitive proof of 4 and 5 and this is where so many problems are occurring with new units. In the Jaguar market in particular there are cheaper new units flooding the market and as a consequence many owners have been experiencing many and varied problems with their brakes, not helped by poor manufacture also.

We take the view that where practical, it is much better to offer non modified components for safety critical components and in the case of our servos, they are dry and wet pressure tested for extra peace of mind and furthermore each one is tested which is more than occurred at Girling. Dry testing means the decay of the vacuum is checked.'

John Wearing also sells reconditioned Girling units, at £195 (exchange), but regards a new Lockheed servo at £180 (outright) as a better bet for most owners, giving the following reasons:

1. 'Why buy a 40 year old unit that's been reconditioned when you can have a new one for the same money?
2. Even correctly reconditioned units sometimes last only two or three years.
3. Our Lockheed units come with all the correct mounting brackets for ease of fitting.
4. No need to pay hefty surcharges on your old unit or postal charges to return it.
5. Any new unit should outlast a reconditioned one.
6. The units I now supply are of the correct boost ration and work every bit as well as Girling.

7. On originality, does it matter unless your car is 100% original and concours? You still have your old unit should you wish to join the rivet counter brigade. Reliability counts most.
8. On the late cars a fluid level warning light was fitted – have you ever wondered why?'

Personally I wouldn't bother trying to rebuild my own unit now, given the high risk of it not working and the alternatives now on the market. Others with greater workshop skills may still prefer to do a home rebuild, given the large cost saving. If there's a moral to this tale, it must be to buy from known Rover parts suppliers with a good, consistent track record.

Tim Pearson ***November 2007***

MY EXPERIENCE OF REPLACING A GEARBOX ON A P4 100

Last summer I decided to renew the badly worn gearbox. The noises had become that strong that I had the idea one day I would need the AA to come home. About ten years ago I did renew the clutch so I was familiar with the work. Because I've never dismantled a gearbox to renew the bearings, I ordered a reconditioned one from a well known supplier in England.

The previous time I had removed the gearbox I had the P4 on four axle stands, and had the engine hang in a workshop crane. This meant I had to lay on my back on the floor when I had to work under the P4. So for every missing tool I had to do some gymnastics on my back to work my body from under the P4.

Since I now have a four post ramp I would this time benefit from it. But how can I 'fix' the engine without being connected to the gearbox? The only option was to make a 'special tool'. After some welding, drilling and painting, the blue 'P4 Engine Stand' was ready. The stand is set on the chassis legs at the back of the engine bay and is held in place by two connection rods fitted to the radiator mounting.

At the rear of the engine the lifting bracket is connected to the engine stand via an adjustable turnbuckle. This turnbuckle has to be adjusted to keep the engine in the 'normal' position when the gearbox is being removed. I fitted a rope as extra support in case the primary turnbuckle would fail (which shouldn't happen because of the size and thus the weight this buckle can support).

From now on it's a straightforward job inside the P4: remove the front bench, carpets, gear change lever and support bracket, cover for gearbox, propeller shaft, wires and under the P4 (standing this time instead of laying!): disconnect gearbox mounting rubber, speedometer cable, clutch shaft and linkages, all bolts which connect the bell housing onto the engine.

Now you can disconnect the gearbox from the engine inside the P4 either by two or four strong hands (my son Edwin helped me the first time) or with help of the workshop crane. The next step was cleaning the gearbox thoroughly.

Because the reconditioned gearbox was without the overdrive unit I had to exchange this unit. In the Workshop Manual you are advised to put the gearbox with the overdrive part in a vice, remove the bolts and lift the gearbox from the overdrive unit. But the gearbox/overdrive is so long and so heavy this is only possible if the vice is at ground level so you can lift the gearbox without breaking your back! So I needed another 'special tool'. This time I made two separate tools, one bolt-on bracket to fit onto the extension casing and one bracket I have welded onto an old rolling chassis stand which is low and steady enough to put the gearbox on.

The next step was flushing the gearbox oil, fitting the bolt-on bracket, lowering the gearbox with help of the workshop crane onto the gearbox stand, bolt the bracket onto the stand and unbolt the gearbox from the extension casing. Now it was easy to lift the gearbox by hand. After making a new gasket the reconditioned gearbox was mounted onto the extension casing. With the gearbox hanging in the workshop crane it was easy now to put the gearbox back in the P4 in position (with of course a little 'struggle' to get the input shaft in place). But……it wasn't that easy: the bell housing couldn't be fitted onto the flywheel housing! Whatever I tried there was still a distance of about a centimetre. Maybe the input shaft of the new gearbox was different to the original one? To check this I took out the gearbox again but the thickness of both shafts was the same! What to do now?

There had to be something wrong with the input shaft so I took the 'old' clutch plate I had renewed some years ago and now I could see what caused the problem. The clutch plate moved smoothly over the old shaft but got stuck on the splines of the reconditioned gearbox. The old shaft looked as new and 'shiny' on the splines whereas the 'reconditioned' one was badly pitted by corrosion. Probably the gearbox had lain in a wet area for many years and was cleaned for reconditioning. It obviously hadn't been used for many, many years.

The only solution to the problem I could think of was exchanging the input shaft too. But I had never dismantled a gearbox so I started with the original one. I also didn't know the sequence which has to be followed: starting at the rear or at the front.

I started carefully at the front by dismantling all the parts to get the input shaft free. Luckily it was a straight forward job with no surprises. Now I had the shaft and bearings on the workbench I could see what had caused the terrible noises in the gearbox: the bearing for the input shaft was badly worn and the bearing for the main shaft had lost a part which was on the bottom of the gearbox!

I now dismantled the input shaft on the reconditioned gearbox and fitted the 'old' one in place. Again with help of the workshop crane I put the gearbox back in position and this time the box went in very smoothly. I was happy again!

After fitting all the other parts I drove the P4 from the ramp and did a test drive. The gear change was not as smooth as I had expected, but using the P4 will solve that, I hope. The absence of the terrible noises was very, very pleasant.

Of course I have related my experiences to the supplier of the reconditioned gearbox. First I wrote him a letter so I could describe my experiences in all ease, and wrote him I would call him after he had received the parcel with the 'wrong' input shaft, which I had sent with the letter.

Two weeks later I made the call. Thc supplier said this was the first time he had heard of my problem. When I ordered the reconditioned gearbox he offered me a bearing set instead, as being much cheaper. Because I had never dismantled a gearbox I preferred a reconditioned one.

Because of all the extra work I had to do, to get a useable gearbox, and because I have now a partly dismantled gearbox, I suggested he would send me a bearing set for free as a sort of compensation, which he agreed to. He also offered me a set of original Rover springs for the gearbox. He hadn't yet checked the 'wrong' shaft due to lack of time.

Some weeks later I received a parcel from him with only another (also badly pitted) shaft with gear, but no bearings, and no note with it. When I tested the clutch plate on this shaft the plate could shift over the length of the splines, so this one seems to be useable. I phoned the supplier to tell him the shaft is OK and asked him why he didn't ship the bearings and springs. He answered he had tested the input shaft I had sent back with several clutch plates and that they all gave no problems. Because the separate shaft he had sent to me was also good, he said that he had done all he could do. I didn't agree because he had promised me the bearing set for all the work I had to do to be able to fit the expensive (!) "reconditioned" gearbox. He still kept saying he had done all he could do.

I was very, very disappointed, which means I will never ever do business with this supplier again!

This story took place in July last year and I have done a lot of joyful driving since. The half dismantled gearbox is still on my workbench and maybe I will buy a set of bearings from another supplier to recondition the box. This will add another experience to my ownership of this beautiful Rover, 'One of Britain's fine cars'.

Alfred van Buuren, Holland — ***May 2008***

ONE CAUSE OF OVERHEATING

I recently had a serious overheating on my 1963 110.

After driving the car for about ten miles and having to stop at a roundabout or crossroads the engine would cut out, and be very difficult to restart.

It was by accident that I discovered the problem. Normally when you turn on the ignition, the petrol pump should work for a few seconds and then stop; mine didn't. In that case with the ignition on remove the rubber tube from the carburettor to the air cleaner elbow assembly. If petrol comes from there, look further. I checked the float needle valve and lever which were okay, the problem turned out to be a leaking brass float containing enough petrol to stop the needle valve from working, ensuring that petrol was being continually pumped from the carburettor into the elbow assembly, thus causing overheating.

I replaced the brass float and the car has been working perfectly ever since.

Richard E Wren — ***July 2009***

DIY RE-UPHOLSTERY – IT CAN BE DONE!

After a very expensive divorce, I needed a cheap everyday car. A Rover 100, rush green in colour, and in average condition was bought, good enough after adjustments to the mechanical side.

So now I could spend my pocket money on it and bring it back to good condition. The green leather on the front bench seat was well worn after the years of driving and falling to pieces. The back seat leathers had been scratched, probably by a very large dog.

So a decision was made to take out the front bench seat and buy two second hand single seats, with better leathers.

The seats bought were in good condition but the leather was as hard as cardboard, so creams were bought to massage the leathers and soften them, over several months. Sadly to no avail.

Whilst spending a week in Bournemouth to visit my many relatives I bought a two day pass to visit the Beaulieu car boot sale, to look for any P4 spares, and found a stall that sold leather at reasonable prices. I explained that I would need enough leathers to cover two front seats and a double rear seat of my P4. I was informed that I would need four cow skins for the job intended.

A skin is only the back of the cow from the neck to the rump, the underside is missing to make it easier to remove the innards, ribs and other parts.

After agreeing a price for the four skins required – dyed green – to be sent to me in London, some weeks later a telephone call informed me that the stall holder could not get the four whole skins, but could supply eight half skins for the same price. I did not mind, so long as all the skins were of the same shade and five feet long minimum.

Sketches were made of the old seats, and measurements taken and recorded for the sewing required.

My sewing machine was not powerful enough to penetrate a single leather with calico backing, let alone to sew two leathers together to form the seven pleats of the seat design.

An industrial sewing machine was bought, which was powerful enough to sew my fingers together if not withdrawn at the last moment.

The horseshoes for the squab and backrest of the front seats were also sewn together inside out to the seven flutings, and including the beading around the edges.

After stretching the complete, new leathers over the seat frames I could not line up the seven flutes of the squabs with those on the backrests, to make it look like one continuous flute as Rover intended. I could not understand why this error had occurred, but it worried me so much I checked out every Jaguar, Range Rover, Rolls-Royce and Bentley. To my amazement they had the same error as I had produced.

Interestingly, a few days later I watched Prime Minister's Question Time, and noticed that the same flaw is present in the hallowed benches of the House of Commons! Those benches are, of course, upholstered by the country's most skilled craftsmen, and use the best leather without any wounds on the animals to work around.

Being satisfied with the two front seats I decided to attempt the large rear one. The 28 flutes were now 'easy peasy' to sew together, although this seat was as wide as the car, and so would need half a cow skin to make a horseshoe for the backrest and squab, with a beading around the edges made from the spare leather around a curtain wire core. The machine was therefore punching through four thicknesses of leather to assemble the horseshoes, beading and side pieces.

The final upholstery task was to tackle the seven armrests.

After adding a new Wilton carpet, the interior regained that lovely smell and ambience that Rover intended.

To think that humans have been sewing animal skins together for thousands of years in order to keep warm. Yet it seems as if there is a secret world of leather working, to which only the privileged could be admitted.

No books could be found on leather working, nor could I find any night school running classes on the subject.

Michael Walkling ***November 2009***

A PERFECT FINISH

Getting a showroom shine on your car is child's play when you know how.

Your classic car is your pride and joy and there's nothing better than taking it out for a spin when the sun's shining. But whether you're driving for fun or on the way to a specialist event, sometimes it takes more than just the car itself to make heads turn. A car that boasts gleaming paintwork, shiny chrome and the sort of deep polish that wouldn't go amiss in the smartest showroom is a joy to behold. That show-stopping finish can be yours with some elbow grease and the right products. Here's a step-by-step guide to getting it right.

Step 1 Cleaning up

First rinse the car with cold water to loosen dirt. Then take two buckets, fill both with cold water and add the recommended amount of professional cleaner to one bucket (washing up liquid is a no-no – designed to strip grease from your dishes, it's bad news for paintwork).

Then using a sponge or wash mitt, clean the car with the soapy water, rinsing your sponge clean in the other bucket between wipes so you don't spread dirt over the exterior. Change the water in the rinsing bucket regularly. Be methodical, thoroughly cleaning each section before moving to the next.

Don't forget your wheels – leave the sponge to one side in favour of a soft bristle brush. Scrub the wheels thoroughly then rinse clean with cold water. Rinse the rest of the car thoroughly with cold water and dry it off using a chamois leather or specialist towel.

Try: Meguiars Nxt Generation Car Wash, Meguiars Soft Wash Gel, Meguiars Ultra Soft Wheel Spoke Brush, Meguiars Versa Angle Tyre Brush, Meguiars Lambs Wool Wash Mitt, Meguiars Natural Chamois and Meguiars Water Magnet Drying Towel.

Tip: Whenever possible, wash your car on a dull day as bright sunlight can dry water quickly, giving an uneven finish.

Step 2 Use a clay bar

This will remove any surface contaminants (such as tree sap and paint overspray) that can bond to your car's paintwork, giving it a gritty feel when you run your fingers over it. It can also dull the finish, making your paintwork appear lacklustre. As well as removing fine particles from the paintwork, using a clay bar means that waxing becomes much easier.

Make sure your clay bar is soft enough to use – it should be easy to mould or roll. If it's not, then leave it in some warm water to make it more malleable. Always use your clay bar with a detailing spray. Spray the paintwork then rub the clay bar across it. The detailing spray acts as a lubricant for the clay so that it slides easily over the surface, but keep it wet so you don't rub the clay over any dry paintwork and damage it.

Always dry off each area as you go – a microfibre cloth is ideal for this. And remember to rub your fingers over a section of paintwork cleaned with the clay bar then do the same with a section you haven't clayed – you'll feel the difference immediately.

Try: Meguiars Quik Clay Starter Kit, Meguiars Quik Detailer and Meguiars Smooth Surface Clay Kit and Meguiars Supreme Shine Microfibre Triple Pack.

Tip: If you drop your clay bar on the ground, leave it. It will have picked up all sorts of dirt and particles and could damage your paintwork. If you have 'butter fingers', divide your bar in two and use one piece at a time.

Step 3 Deal with scratches and imperfections

Any serious scratches should be dealt with by the professionals to ensure a high quality finish, as taking the paintwork down to the level of the scratch calls for experienced hands. But fine surface scratches, swirl marks, oxidation and blemishes can be tackled at home.

Take your scratch liquid and apply following the instructions. For paintwork stains and oxidation marks use a paint cleaning product. Your product will fill in light scratches and abrasions, which well help prevent further wear and tear to your paintwork. As well as restoring damaged, dull paintwork, and enhancing its colour and clarity without scouring it, a paint cleaner or scratch liquid will also leave your car's surface perfectly prepped for waxing and polishing.

Try: Meguiars Deep Crystal Paint Cleaner, Meguiars Scratch X.

Tip: Always remove your scratch product or paint cleaner before it dries, so use it on small areas and remove thoroughly before applying more elsewhere.

Step 4 Polish up

Some confusion exists over polishes and waxes. The difference between them is simple, a pure polish is a blend of oils that the paint absorbs and this creates a deep gloss. But the gloss is all you get – a polish offers no protection for your paintwork. An application of a wax product on other hand, will seal in that deep glossy sheen that a polish gives and it will protect your paintwork to boot. See Step 5 for more details.

For the best results, follow the instructions on the pack carefully. But generally you'll be on the right lines if you apply a small amount of polish to an applicator pad (or pure terry cloth towel), then working over one section at a time, rub thoroughly into the surface. Wipe dry with a clean microfibre cloth or pure terry cloth towel.

Try: Meguiars Deep Crystal Polish.

Tip: Make sure that your car is washed, cool and parked up somewhere nice and shady when you're using a polish.

Step 5 Wax works

Your paintwork is clean, prepared and ready to wax and buff to a high shine. Waxing will help protect your paintwork from the elements and give it that glossy finish you're after. Take your pick from a vast array of products – these days there's everything from a traditional tin of wax paste to high-tech sprays and liquids.

Always follow the instructions carefully to get the best results. However as a rule of thumb, leave waxes to dry before starting to buff. The best way to do this is to apply wax to the whole car then begin buffing and polishing where you first applied the wax and work your way methodically over the surfaces.

Try: Meguiars Tech Wax 2.0, Meguiars Cleaner Wax Liquid, Meguiars Deep Crystal Carnauba Wax, Meguiars Cleaner Wax Paste.

Tip: Rub off any polish that gets on your glass at once. If you get it on your trim, use a grease or wax remover, following the instructions carefully.

Step 6 The details

Gleaming chrome, dark tyres and smart trim will set off all the hard work you've put into getting a deep gloss on your paintwork. Revitalise tired-looking bumpers and trim with a spray to restore colour and sheen as well as create a protective shield against dirt, moisture, grease and weathering.

Don't let scruffy tyres ruin the look of you car, use a tyre dressing, gel to smarten up stained tyres. Chrome and metal should gleam to match the paintwork – use a metal polish on tarnished or pitted metalwork to bring it back to life. If you have alloy wheels, once they're clean think about using a wheel seal to help them stay that way and make cleaning easier in the future.

Don't neglect your windscreen and mirrors – they'll also benefit from a polish, use a glass cleaner and buff up for streak-free finish.

Try: Meguiars All Metal Polish, Auto Glym Bumper Black, Meguiars Nxt

Generation Glass Cleaner, Auto Glym Alloy Wheel Seal, Meguiars Quik Detailer.

Tip: Maintain that pristine look by keeping a bottle of spray detailer to hand. Just mist on and wipe off with a microfibre cloth to remove any fresh dirt or contaminants on your paintwork.

January 2010

REAR SCREEN SEAL – DISPEL THE MYTH

Looking at the rear screen seal on my Rover I decided it needed renewing as it was perished. Mentioning this to an ex-Rover man he strongly advised against this: 'Takes days,' he said. This was followed by Classic Car Weekly in a feature for P4 Rovers. They quoted: 'This is a tricky job as replacing it can take 24 hours with labour charges at £24 per hour typical, and is not a DIY job.' No matter who I talked to it was a horror story. So having been warned, I decided to have a go.

Armed with the new seal I set to battle with an open mind. Removing rear seats, interior lights and trim panels is not difficult and the rear 'parcel shelf' came out easily. Then removing all the window clamps and spacers, I was amazed to be removing the rear screen in its frame in about an hour.

There's the fiddly bit of removing the old seal and fitting the new one.

The groove for the seal must be meticulously clean and the new seal 'fed' in with a blunt screwdriver and much Fairy Liquid for lubrication. For this operation allow two hours to do it properly. The screen frame and new seal is now ready to replace.

It was at this stage I put the frame to one side and did some remedial paintwork. At some time the car has had paintwork and the rear screen masked out, unfortunately this left a step or paint line and the new seal may not marry up with it.

As with any paintwork it is time-consuming to achieve the necessary finish a quality car demands.

Having completed this I resumed battle with the screen. My plan was to fit the screen and frame 'dry' – that is to apply the sealant when the frame is held by a few clamps. This proved successful as there is a gap between the frame and the body on the inside and a thin nozzle on the sealant gun does the outside.

Then all the remaining clamps and spacers were fitted and tightened. The fiddling with clamps and refitting of the interior always takes longer than pulling it all apart but overall this job takes around six hours. Granted you need application, determination and a few swear words but it's not the horror story it is made out to be. In fact I've done both front and rear screens single-handed and I know which one I'd sooner do: the rear one.

Dave Horrocks ***March 2010***

(Editor's note: This seems a DIY-able repair if the owner aims only to replace the outer rubber seal, which is fitted between the metal frame and the car body. The inner seal, fitted between each piece of glass and the frame, can be a much tougher beast to control.)

ANTIFREEZE

In the 2009 Yearbook we highlighted the dangers – both to cars and to human and animal health – of various types of antifreeze. The basic message was that the best types of antifreeze for modern cars may be far from suitable for classic cars.

This issue is now recognised by the Federation of British Historic Vehicle Clubs, which has circulated the following information amongst its member clubs:

'Bluecol and Blue Star are well known brand names and both of these are declared suitable for classic cars.' Perhaps we should clarify that we were referring to the traditional blue coloured Bluecol – but the company also sell a red coloured Organic Acid Technology (OAT) product suitable only for modern cars, not classics. Even more confusingly, there is also Bluecol U which is marketed as a universal top up and not an antifreeze product with which you would fill the whole tank. The manufacturer has assured us that this is suitable for historic vehicles.

It has also been brought to our attention that Halford's sell a blue-coloured 'Advanced' antifreeze which has a label containing the phrase: 'Older vehicles can further benefit…' but on further examination it was discovered that this product does indeed contain OAT and therefore cannot be recommended for historic engines.

Our postbag has also been swelled by correspondence relating to the extremely poisonous nature of ethylene glycol, indeed the Cats' Protection League have gone so far as to start an on-line petition to highlight the danger to small animals accidentally ingesting tiny quantities of the product.

Propylene glycol is much safer and one of our new trade supporters, AAA Solutions Ltd, is about to launch a propylene glycol based antifreeze specifically aimed at historic vehicles.

It does remain a rather confused picture, but the important facts to remember for historic vehicle owners are: use only Inorganic Additive Technology (IAT) products according to the manufacturers' instructions and take great care with any liquid containing ethylenc glycol.

Coolant Facts

1. Engines develop heat.
2. Coolant is needed 24/7 to cool the engine.

3. Water is the best heat transfer fluid, but it is highly corrosive to metals, freezes at too high a temperature, and boils at too low a temperature. Distilled and de-ionised water is the most corrosive, while hard tap water develops limescale.
4. Glycol is added to lower the freezing point (Antifreeze) and raises the summer boiling point (which is helped significantly by having a pressurised system). There are three types of Glycol:
   - Ethylene Glycol (petroleum based) which is TOXIC and a poison, metabolises into oxalic acid which causes kidney failure in animals and humans, and main bearing failure in engines.
   - Propylene Glycol (petroleum based) which has low toxicity, does not harm main bearings with the right inhibitor package, and metabolises into the citric cycle. It is rated as GRAS (generally regarded as safe) and is used in food and cosmetics.
   - Propylene Glycol (sustainable) with all the PG advantages but made from glycerin, a by-product of green bio-diesel.
5. A proprietary inhibitor package is added to the mix to deter corrosion, and deal with other important factors. The percentage of water to the mix and the type of water is a customer problem with a concentrate. The inhibitor package is the differentiator between brands which can be very good for all vehicles (like AAA Classic Coolant), no good for heavy use (SCA), or very bad for vehicles manufactured before 2002 (Longlife, OAT, HOAT, NOAT). Some need changing every two years (IAT), or five years (Longlife, OAT, HOAT, NOAT).
6. All variations of Antifreeze/Coolant have passed the standards (BS6580, ASTM, AFNOR). Colours have been added to differentiate the Inhibitor Packages only to create more confusion. The end user cannot rely on colour or the Standard to make the right choice.
7. AAA Classic Coolant make it simple: a pre-mix with water, suitable for all vehicles, cools better, lasts longer. With its Quality Installation and Maintenance Procedure (QUIMP) it will last between major engine overhauls or the life of the vehicle. It lowers servicing costs, antifreeze disposal costs, and improves reliability for all vehicle owners. Fleet Operators have realised significant cost savings.

8. Cost Comparisons:

Old Blue Antifreeze Pre-mix £17.99 lasts two years. £9 per year plus disposal costs. OAT Long-life Pre-mix £19.99 last five years. £4 per year plus disposal costs. AAA Classic Coolant Pre-mix £19.99 10+ years life. £2 per year.

AAA Solutions Ltd ***September 2010***

OILS FOR YOUR P4

I am frequently asked by our membership for advice about the best oils to use in our P4s.

Starting with the engine, I must say that I believe the P4 being that it was a 1930s/1940s design does not require a modern fancy oil. Many oil companies, in my opinion, are cashing in on the classic car market and though their oil is superb, it is being sold in beautifully produced tins at about twenty pounds per gallon. Most P4 engines burn a bit of oil, and even more they are mostly incontinent and spill large amounts on roads, driveways and garage floors. So I think there is nothing to be gained by spending and indeed wasting money on over marketed oils for the engine. I use, and have done for many years, Asda 20/50 multigrade oil which is currently available for six pounds for 4.5 litres. So for twelve pounds you can change the Rover's oil perhaps a little more frequently than the recommended 3,000 miles. This oil is produced by Tetrosyl and is just a basic oil with no detergents or additives, making it ideal for our Rovers. I have covered 200,000 miles in my 95 using it.

Moving on to the gearbox, I use an 80/90 EP oil in this unit for P4s with and without overdrive, and the make I use is Carlube, again produced by Tetrosyl. This oil is also ideal for the rear axle.

The steering box, steering idler and swivel pins require a SAE 140 oil and I use an oil from the Millers Classic Gear Oils range. I obtain mine from a very useful company called Frost Auto Restoration Techniques Ltd who stock it all the time. It can be difficult to find at accessory shops.

Brake fluid should be a DOT4 configuration and is mostly synthetic these days. Nothing wrong with that, but remember it can still take paint off!

Synthetic is fine, but must not be confused with silicone which can be lethal if mixed with synthetic brake fluid. The only time silicone can be used is when the entire brake system has been rebuilt and has never had ordinary brake fluid in the system.

Lastly, anti-freeze. The ideal anti-freeze to use is the one advertised in Overdrive by Classic Coolant, this is superb and very safe in the P4 engine and although more expensive, will last a lot longer in the system. Having said this, most good quality anti-freezes are fine, particularly Bluecol.

Stan Johnstone ***January 2011***

(Editor's note: A widely-held myth is that SAE 140 went out of production in the 1970s. By coincidence, the day before Stan sent me this article I'd bought a litre of Miller's SAE 140 on eBay for about £12 including P&P from Specialised Motor Components Ltd, 291-301 Parkstone Road, Poole, Dorset BH14 9DZ Tel. 01202 744244. On eBay their 'handle' is 'esemsea'. The Proprietor's son Murray Dick confirmed that if you mention this magazine when buying over the phone or on eBay, he'll send the oil first class on the same day. He may also be willing to offer discounts for anyone buying in bulk.)

LEATHER UPHOLSTERY REPAIRS

Following on from a discussion about repairing leather in another thread, I've made a start at bringing my seats back to life.

I've been in contact with a company called Furniture Clinic about how I should approach this job. I'm sure other people offer a similar service, but I can say that these people have been very helpful in the advice they have given me.

I'm far from finishing all my seats yet, but thought I'd share some pictures of the now almost complete drivers side base. I started with this as it was by far the most delapidated area of leather on the car. These photographs show cracked and split leather which generally looks a mess, although most it is holding together apart from a 2" split:

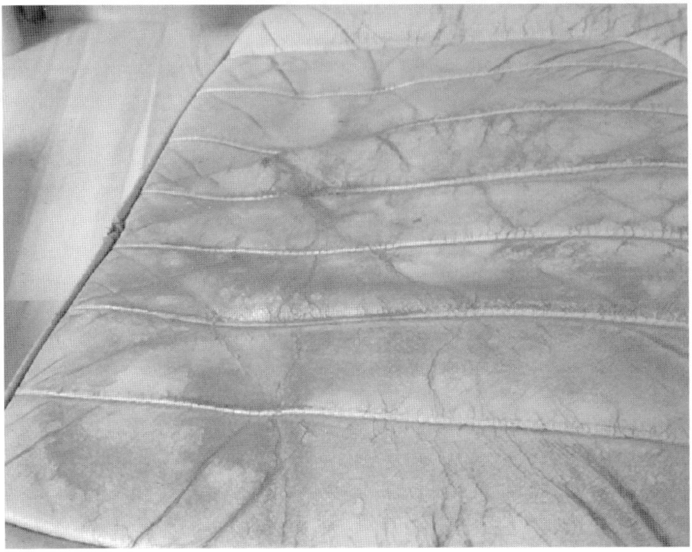

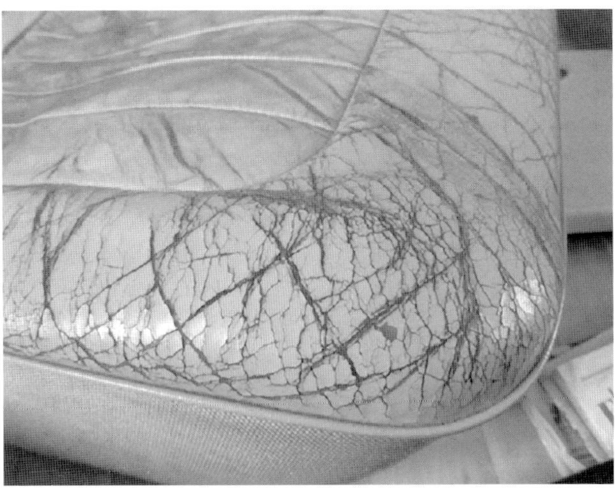

So this is what I ordered:

It consists of a colourant kit, a leather repair kit, leather binder and flexfil. Costed in all about £150, which sounds expensive, but compared to the cost of retrimming the car, is a far more economical restoration method and preserves the original car materials...

Okay I started my using the solvent, then alcohol based cleaners which take the top finish away from the leather. The result alone after this process was quite good.

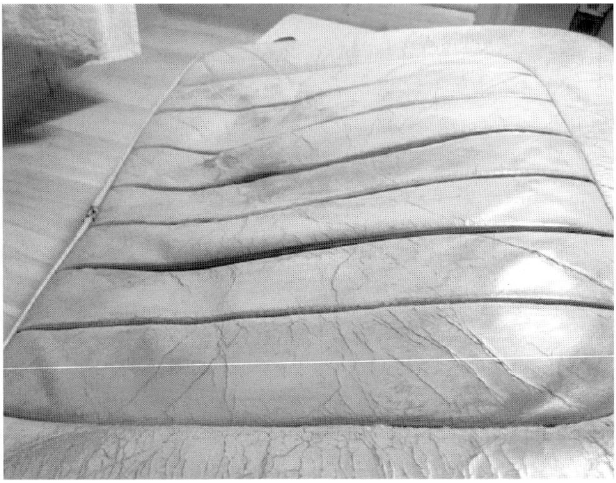

I then had to apply a few coats of leather binder to the seat which is supposed to strengthen the fibres.

Once this was done it was time to tackle the split. This was a case of pushing some material behind the split and applying the supplied glue.

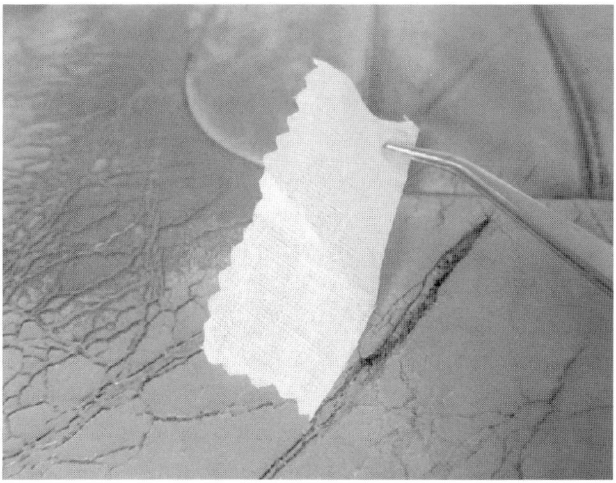

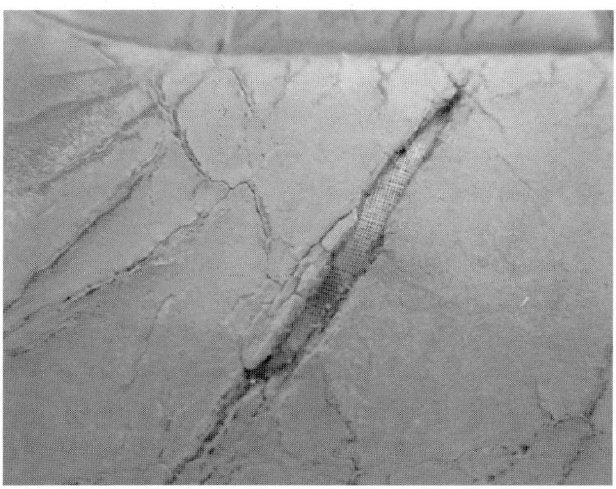

Once the glue has dried, thin applications of the flexfil was applied to all the cracked leather and the repair. Applying about 2 coats a day this process took about a week to get a good smooth result. I could have taken even longer for a totaly smooth finish, but that isn't the result I'm after.

Photo showing initial application of flexfil:

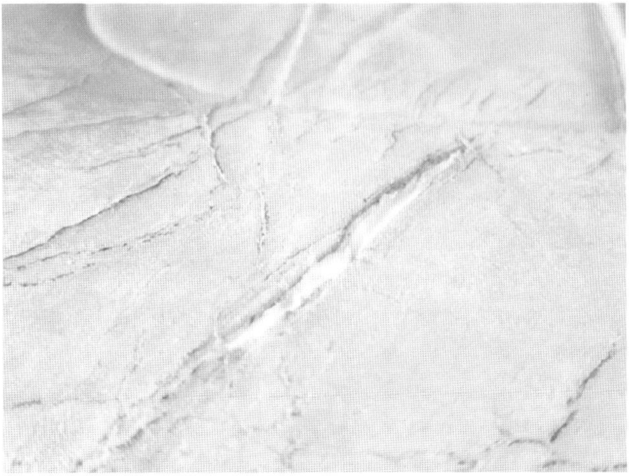

And after a few more thin applications when it was almost smooth:

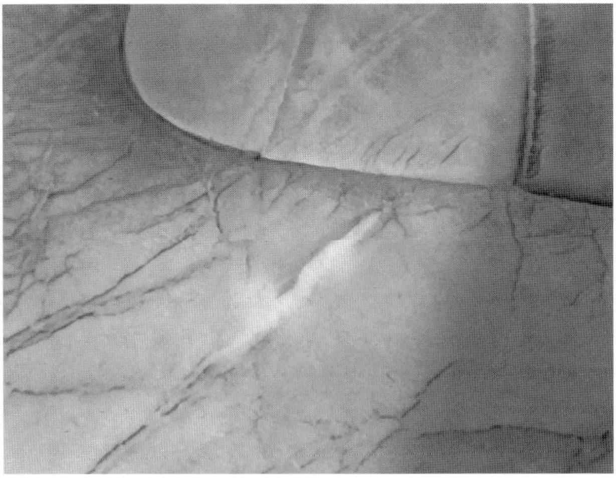

Now it was time to apply the colour. I originally send an armrest to FC when I placed the order so they could match the colour. The initial coat is applied with a sponge.

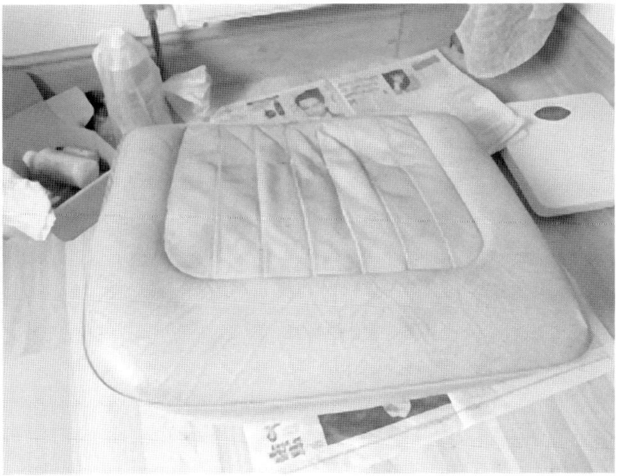

Subsequent coats are applied with an airbrush until a uniform colour is achieved. Following that top coats of clear finsh are applied (Satin finish) resulting in this:

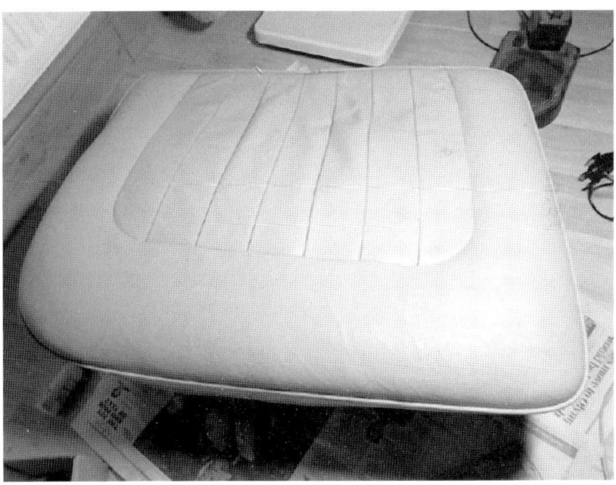

I've now got to wait a week before applying a leather feeder/ preserver which will need to be applied every 3-6 months to keep it in good condition.

This is the first completed item of leather. I have a production line of various other chair parts, arm rests etc which is likely to take me another month to complete. I have to say it's not a fast process (flexfil applications take ages if you can only apply it a couple of times a day), but the end result I have to say is a huge improvement. Worth the effort I think!

Mark Baird ***March 2011***

HANDBRAKE RATCHETS

Reference Part No. 234055. For those of you who search the Internet for spare parts for your P4, you may have come across a website selling Rover spares that makes references to my handbrake ratchets (quadrants) having teeth cut at the incorrect angle. This might be partially true, they are different to the original Rover one and for good reason. When we had this item remanufactured, we changed the angle slightly, this allows the pawl to sit further into the teeth on the ratchet, thus giving both components a far longer life span and a greater safety margin. So while we have taken note of the comments, you will now understand our reason for this small but improved modification, and if you talk to any long term P4 owner, who owned their car when it was relatively new, they may tell you a tale of chasing the P4 down a street, as it had slipped on the handbrake and ran off down the road, many ending up with catastrophic results.

We also know they work better, as many have been built up and fitted in our own workshop or sent out to customers as a built up unit for them to install on their car. The first production batch was made for us locally and cut using a computer controlled CNC machine, which is far more accurate than old fashioned engineering methods that some companies still prefer to use. At the time of making these, I can recall new ones being nigh on impossible to find, and used ones making similar prices to our brand new ones – if you could find a used one with any teeth left on it. To date we have sold just over a hundred units of these and have had a second batch remade, all without one complaint (sorry, just one, it is on his website). I think the results speak volumes.

Whilst on the subject of handbrakes, we also have the following parts reproduced exclusively for ourselves, these are the pawl or catch part no. 219834, the two aluminium wedges which sit between the ratchet and front outrigger parts no. 263974 and 265287 and the pull off spring 238966. Sadly though these three items have been made in a foreign land (to us) – deepest Yorkshire (Although a Yorkshireman I'm sayin' nowt – Ed). In our own workshop, we remake the fulcrum pin part no. 265286, the ratchet cover 263476 and the bracket that holds it in place 232902. We did cheat slightly on this item – we used ¼ UNF for fixings instead of the ¼ BSF originally used. Rivet counters look away – no one will notice this very slight error I hope. We also stock the little bush part no. 01379, which wears quite badly and can cause damage to both the ratchet and pawl caused by excessive sideways movement.

I will also be remanufacturing the little wire link for the return spring part no. 263518 as we find these are often missing and made up from a bent piece of wire. For those of you wanting to improve their handbrakes, the previously mentioned parts will certainly make the lever tighter and more precise than it has felt since the car was new. To finish the job off and, of course, keep the draught out of your right hand trouser leg, we also have the rubber floor grommet part no. 231015. This item has normally perished long ago.

Finally, we can also re-chrome in triple plate your shepherd's crook lever without damaging the little black lever grip, or can offer the whole assembly rebuilt for you, this would include all previously mentioned parts and re-chroming of your lever.

I hope you are now correctly informed from our knowledge of the P4. I will be displaying assembled ratchets at the National Rally.

John Wearing ***May 2011***

THE VOLTAGE REGULATOR – HOW TO SPOT A FAULTY ONE

Sadly my Rover 'Bertha' and I did not get as far as we would have liked on Sunday 8th May. We were on our way to the National and had crossed the river at Dartford, and were approaching one of the two roundabouts to get us back onto the M25 at Thurrock Services.

The traffic lights stopped us, and just at that point I looked at the instruments and almost panicked as I saw the ammeter showing maximum discharge. We had just stopped and I reached over to switch my battery isolating switch to off. I quickly risked jumping out – road clear at that moment! There was no immediate smell of drastic overheating or any other signs of any electrical fault. Decision time. We could not afford to stay where we were, as the risk of being hit was too great. Could I afford to risk switching everything back on? I decided the risk in staying there was even greater, so started up and tried to get out of there quickly and back to the Services. We made it!

I then shut everything down and did some investigations whilst gathering my thoughts. Something in my mind recalled a similar incident last year after the Teddy Bears' Picnic! No sign of heat or damage but the symptoms persisted. Could the voltage regulator be faulty? I wondered whether this was failing and possibly forcing the dynamo from generator to motor mode! There was little more I could do at the time, and I decided the problem was too great to deal with where we were. Decision sadly but safely made: no National and abort the trip.

Bertha is insured with Footman James, and my memory told me that they provided a recovery service, but I did not know my policy number although their number was on my mobile phone, so I called it. One ring later, a young lady answered the phone and was most pleasant telling me not worry, just give her some information about us and she would do the rest. A few moments later she was saying that she had located us on the system and was then asking more questions, what was the problem? Could it be fixed on site? What did we need? Where were we? And so on. Her computer had located our position, so we left the rest to her, confident that she would get things moving.

About five minutes later the mobile phone was bursting into action, with a text message that Dynes Emergency Breakdown of Crayford had been alerted of our predicament and would be with us in about 45 minutes. About twenty minutes or so later I saw their truck coming across the car park, I attracted the driver's attention, and he was soon in front of us with orange lights flashing to warn people to keep away. He had a short discussion with me and a quick look at things, before deciding the same as I had. He then asked me where we wanted to go before getting Bertha on to the back of yellow truck.

We were on our way home and for the first time since 1996 I was not going to make the P4 National, so rang Stan to make him aware.

Two days later I decided to investigate, and decided that the voltage regulator had come back to haunt me. Helpful phone calls followed, firstly to Stan and then to John Wearing. Decisions were made all round with fingers pointing at the voltage regulator. John got a new one shipped out that day by courier. This has now been fitted to Bertha after many labels being fitted and with a lot of caution. Hey presto – things were looking good! After a test drive, confidence was emerging!

Last Sunday was the Teddy Bears' Picnic at Bentley Wildfowl, we attended as you would expect us to. Bertha is now back on form!

You can see the usual team playing their part. Stan for remote advice, John Wearing to provide our parts with a speedy and efficient delivery with ACE and of course Footman James coming to the rescue via their contractor Dynes of Crayford.

Now to complete things, over to Stan's at the beginning of June for her usual annual service and set up for another year, all being well.

John Leng ***July 2011***

CHANGING YOUR P4 CHASSIS AT HOME – IT CAN BE DONE!

I have owned my P4 100 for nearly five years and as it is a 'round the clock plus model'. I have carried out a fair bit of work to the car. It has never been off the road for any length of time and I normally drive it about 1,500 miles a year. I always kept it inside. With limited space of a single sixteen by eight foot garage and a short driveway all car work has had to be done in the open. It was always important therefore that any major project was carried out during dry weather.

Initially I noticed that the offside rear of the chassis was rather thin in places and in fact during a full road spring change early in my ownership I accidentally pushed a screwdriver straight through it. It was repaired, but later I observed more areas that were of concern, and at that point I realised that to give the car a new lease of life it would have to be changed.

I obtained a suitable chassis in good order with no repairs save a pair of new front outriggers. I stripped, sanded and painted the planned replacement during the summer months. Later that year as winter approached I wrapped it in a large tarpaulin, waiting for suitably dry weather the following year so the work could begin.

It had struck me as quite a daunting task and most people I spoke to thought it would be too difficult without a garage lift at the least. In fact I got a quote from my local MG Rover garage based on me doing all the hard work first of removing the engine, transmission, axle and suspension. The quote still came in at over £2000!

The usual way to go about this work is to find a way to lift the body off the chassis then strip it of all running gear and suspension, build the replacement then lower the body onto it and bolt up. Easy. Except that at home I certainly don't possess anything capable of lifting the body high into the air to do this.

My alternative method was to strip the chassis whilst still attached to the car, thereby making it fairly low profile and light, then jacking up the whole assembly high enough to give sufficient clearance to lower the chassis from the body.

(Light is a relative term as the replacement proved to be far from light when a friend and I had to lift and carry it from the back garden to the front of the house where the car was!)

I calculated that the sills would need to be raised to a height of 64cm (25 inches) from the ground to give sufficient clearance, which would enable the chassis to be pulled out rearwards (on rollers) beneath the car.

So far so good. But how to support the body in such a way as not to obstruct the lowering of the chassis and its removal? Axle stands intrude under the car so something else was needed. I had planned that some kind of wooden supports could be constructed and these would be placed below the sills on each side supporting the A, B and C posts. Wood is a very strong material and if properly constructed should easily the support the weight of the non-stripped body shell.

I decided to use standard wall stud timber due to it being a finished product, smooth and available in suitable lengths. I arranged the pieces using wood joints and triangles for strength with further bracing from steel plates.

Outriggers further supported the two sides. These prevented lateral movement when load was applied. Finally the two sides were attached together by two cross members just before the body/ chassis assembly was lowered onto them.

Lifting the car high enough to do this was done by an engine crane at the front and two high lift trolley jacks with some stout timber at the rear.

With the car now resting on the wood supports and the chassis still with jacks taking its weight, the bolts holding the two together could be removed. There are five body mounts each side of the car. Three in each sill, one at the extreme rear and one mounted to the bulkhead. It should be remembered that there are also the front seat mountings to unbolt (they attach to chassis brackets) and the two forward brackets securing the front inner wings to the chassis. Also if not already removed, the support frame to the radiator is also attached to the chassis. In my case all the mechanicals had already been removed from the car so this was not an issue.

The steering box has to be removed and it is easier when the engine is out as it can be pulled forward (not forgetting to take the steering wheel and switches off first) rather than the manual's method of removing from inside the car. Also the brake and clutch pedals must be released via pinch bolts and pulled up through the car. Other important items for removal include the brake pipe from the master cylinder to the reservoir and front brake hoses if the suspension is in place. The handbrake lever and battery cable as it is attached to both the chassis and bulkhead. Check any wiring that may snag when the separation takes place.

So having done all this and a final check all around the car I took a deep breath and started to lower the engine crane attached to the front cross member. There was a bit of creaking then the chassis started to lower. I continued with the rear trolley jacks checking all the while that all was going to plan. It was!

Once the chassis was resting on the rollers I had positioned below I removed the jacks, released the engine crane and pulled the chassis rearwards clear of the car. It was soon apparent that I had made the right decision. The top surface of the rear offside was peppered with holes and the nearside was not much better. Of course these areas cannot normally be seen but the rest of the steel had been repaired several times before.

It was several days before the new chassis was ready to fit. When the time came it proved to be straightforward enough and positioning the mounting shims and bolts presented no major problems. It is a good idea to check the dimensions given in the Rover manual for the mountings as they can vary from one chassis to another, particularly if outriggers have been replaced.

As the manuals say, reassembly is the reversal of the above! And so it was, although during the couple of months or so that the car was stripped I had rebuilt the engine and suspension and repaired the rear body, renewing the mounts and boot floor corners but that's another story!

Peter Osborne ***September 2011***

GUIDE TO SEVICING THE BRITISH JAEGER CLOCK

Whilst I'm not an expert on the subject of electric clocks (long case clocks are my area), I've successfully brought the clock in my P4 back to life, so thought I'd share some information on the subject.

My 1958 105S is fitted with the British Jaeger type clock which is what I've based this article on. To service the clock it needs to be removed from the dashboard. It is held in place with two screws which need to be removed and the U shaped bracket can be slid from the clock casing. Before the movement can be pushed out of the front of the dashboard, the adjustment shaft will need to be unscrewed and withdrawn. The electrical bullet connectors will also need to be disconnected.

Once the clock is out of the car, before attempting any more work, you will need a clean and well lit work area. Basically it's an ideal activity to carry out whilst it's cold outside!

To complete the task you will need:

- A small electrical screwdriver.
- Clock oil (don't be tempted to use anything else)
- Very thin wire to apply clock oil
- A 12Volt DC power supply ideally adapted so it has 5mm bullet connectors. This allows for the clock to be tested easily.
- Good eyesight and a steady hand.

To remove the movement from the case:

- Remove the black cable (+ve earth) from the case
- Pull the rubber bulb holder from the back with the red wire.
- Twist the bezel off and remove the clock face taking note of the order of the parts.

- Lastly remove the two remaining screws (make sure you don't touch the adjustment screw in the middle of the casing) and the movement should come away from the casing.
- The purple wire can then be removed from the movement and the wire left with the case.

Upon inspecting the movement its clear that it's very simple in design, but well made, so there isn't much to go wrong as long as its maintained (clocks should be oiled every three years or so).

It is driven by an electromagnetic coil which encourages the balance wheel to oscillate. The clock doesn't appear to be self starting, but the method of starting the clock is by pressing in the adjustment shaft which moves the balance wheel to one side, where upon releasing it will recoil the balance spring and start the clock. There is an electrical connection which is made when the adjustment shaft is pressed.

If you give the balance wheel a gentle push you should be able to see the escapement movement operate for a few cycles, which will be a good indication of a potentially healthy clock. This part of the movement is jewelled it should be quite long lasting.

The main weak point of the operation of the escapement movement is the electrical contact between a pin on the balance wheel and a sprung arm attached to the middle clock plate. Over time this can become damaged by minor arcing or sparking. Unfortunately, it's incredibly delicate so if it needs cleaning up take great care. I'm going to examine if there are any potential solutions to this and if there is a good improvement I'll write another article. There are two solutions to this sparking which can be suppressed by a capacitor across the escapement contacts, or a diode across the inductance coil. I'll spare you the details on this now and focus on oiling the clock.

Provided that the rubber bulb holder is normally left in place, there isn't really any reason for the clock movement to become dirty inside, so it shouldn't need much dismantling to service it. However if it's very dirty inside or has been incorrectly oiled it will need cleaning completely. If someone has squirted the inside of the movement with WD40 or similar this is a good reason for a full service and clean which is a much more involved task. I did however remove the back plate to inspect and photograph:

When oiling the clock, you need to be careful not to over lubricate it. Too much oil will attract dirt and gradually make an abrasive paste which will wear the pivots and pivot holes. You need to use a high grade clock or watch oil.

The oil should be applied with a very thin wire. Get a small drop on the end and pop it on the pivot hole into the oil sink. Repeat for all pivots. Remove any surplus oil that gets dropped with tissue paper. That's all that needs oiling, don't be tempted to oil the clock wheels (gears) as it's not needed and will cause damage to the movement.

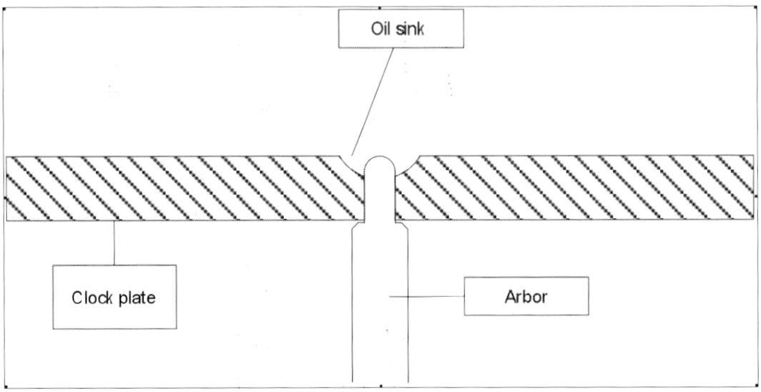

I would not oil the jewelled pivots of the balance wheel as to do this properly, the balance wheel needs to be removed and the pivot cleaned out. You also run the risk of getting oil on the balance spring which can cause poor running.

It is probably easier to test the clock using a 12Volt supply if you have access to one. It can be initially tested with the movement out of the case, but don't leave it exposed to dust for too long.

If you are interested in reading up on clock maintenance further the following books may be of interest:

- The Clock Repairer's Handbook by Laurie Penman (A good general introduction to the subject which isn't too dry!)
- The Science of Clocks and Watches by A L Rawlings (for anyone who wants to jump in at the deep and really understand the physics and mechanics)

Mark Baird ***November 2011***

(Editor's note: should the clock need more extensive work to get it running again, visit the following advice on the Guild's on-line Forum: http://www.roverp4dg.org.uk/phpBB2/viewtopic.php?t=909&postdays=0&postorder=asc&highlight=clock&start=0)

SCREEN SEAL RENOVATION – IN SITU

I have a 1963 P4 110 in the south of Spain and I thought I would share a discovery I made while restoring the said car. It concerns the rear screen rubber which was rather the worse for wear and quite perished in some parts. I will describe the treatment I gave it, and which returned an almost as-new looking result:

1) Clean rubber thoroughly with a coarse cloth or scrubber using cellulose thinners, taking care not to damage surrounding paintwork. This removes the loose and oxidised outer layer of rubber.
2) Carefully mask both the outer and inner edges of the screen rubber, making sure the masking tape follows the original outline.
3) Mix about 50cc of black urethane sealant (Sikaflex for example) with cellulose thinners until it is liquid enough to be applied with a soft 1/2 inch paintbrush. It has a slightly thixotropic consistency so will not drip all over the place. Paint this mixture on the rubber, applying it thickly to the most eroded parts and allowing it to cure until it is touch dry. The time depends on temperature, humidity etc. so you need to do fairly frequent 'finger tests'. When you can safely run your finger across the surface, remove any brush marks by gently smoothing the surface with the finger, shaping any bumps etc. to conform to the original outline. This is a fairly delicate operation but not as difficult as it sounds. Allow to cure for at least 48 hours.
4) With great care, use a scalpel or razor blade to cut the overpainted masking tape to make a clean edge and carefully remove the tape, being sure not to stretch any uncut new rubber, which could go out of shape.

And that's it. I must confess I was very surprised at the result and hope this tip may be of use to fellow members.

Nick Wadham *November 2011*

WHERE'S MY BRAKE FLUID GOING?

I am sure I do not have to remind you of the importance of keeping an eye on the brake fluid level in your P4 brake system. Every week you should be looking at the fluid level if you use your P4 regularly, if not at least once a month. With some problems you can be losing fluid without even driving the car. If the level has dropped in any way, I suggest you make the following checks:

1. Look under the car beneath where you would be sitting and you will see the brake master cylinder – 105R under the bonnet. If it is wet around the unit, you will have to think about a new master cylinder or at least to rebuild. Generally speaking this will be governed by how much the bore of the cylinder has corroded, a new seal will not seal for very long if this is the case, so a new one is on the cards.
2. Remove each wheel, one at a time, and remove the brake drum to see if the brake slave cylinders are nice and dry. On a late P4, you will soon see if a brake calliper is leaking fluid.
3. Everything dry? Well, short of a brake pipe failure, it is got to be the brake servo unit itself. Servos do not leak, but they can very craftily drink the brake fluid by allowing it to enter the vacuum pipe and get burnt through the inlet manifold in the bores of the engine. This is the most common culprit for the loss of brake fluid. Again due to their age, they are rarely up to being renovated because the bores in the aluminium corrode and a new seal kit, although readily available, can turn out to be a waste of money – because of their vital role in stopping the vehicle, it is not worth the risk. A new unit is the order of the day. Another thing to think about regarding the braking system is the age of the three flexible brake hoses. Not so much that they may leak but more the fact that brake hoses tend to get smaller inside, a bit like arteries in our body. So when the brake pedal is pushed, servo or not, the push has to be a lot harder to get the fluid through the restricted hoses. More importantly, the fluid cannot always return back through the hose as quickly as it should, and therefore can lead to binding brakes. For what they cost, it is worth changing them. Even back in the 1960s, Girling were recommending that they are renewed every 40,000 miles.

4. The other point is that brake fluid does not last forever. Being hydroscopic it absorbs moisture and therefore corrodes everything it touches. DOT 4 is ideal for our Rovers, but no way should it be mixed with silicone brake fluid. If the two types are mixed they can cause untold damage, by causing rubber components to swell up and seize.

Stan Johnstone ***March 2012***

YOUR ROVER'S STEERING AND THE MOT TEST

On 1st January this year, new regulations were introduced for the MoT Test, some of which will directly affect our cars, especially concerning steering boxes and rubber covers on steering joints etc.

The new regulations state that there should be no oil leaks from steering boxes – I presume this means steering idlers as well – and the steering must be checked for smoothness of operation.

Unfortunately, most, if not all P4s (and P3s) suffer from leaky steering boxes. The biggest cause of leaks on P4s is the lower oil seal (item 5 in the Parts Catalogue). It is not really an oil seal, it is actually a pure rubber `O` ring that over the last 50 odd years has gone hard and will no longer seal on the rocker shaft. It is possible, but very, very difficult to replace the `O` ring in situ (the proper way is to remove the end cap and press the end cap back on). The difficulty is that there is a thin steel washer, (item 6 in the Parts Catalogue) that fits above the `O` ring, the purpose of this washer is to compress the `O` ring to make it a tight fit on the rocker shaft. I have known several owners who have spent many, many hours replacing the `O` ring only to find that the leak is just as bad after all of their hard work, and bruised knuckles!!!!

Another reason for oil leaks is very badly worn rocker shafts and bushes, especially the lower part. P4s are very heavy, and the loads taken by the steering box and joints are immense, especially as more and more owners use radial tyres.

Another area of concern is stiffness and 'notchiness` in steering boxes. This is usually caused by either over tightening of the top adjusting screw, or simply old age. The worm screw thrust bearings (items 21,22 and 23 in the Parts Catalogue) always wear very, very badly, especially on the lower half of item 21 due to incorrect adjustment putting tremendous pressure on the bearing, and causing it to fail and the steering becoming `notchy`. I've seen many of these bearings with huge `gouges` out of them – it's a wonder they work at all!!!!.

Over the last few years I have reconditioned many P3 and P4 steering boxes – including left hand drive – and steering idlers, and I can now offer members the opportunity to have their boxes and idlers fully reconditioned, on an exchange basis.

The steering boxes are stripped, cleaned and inspected. The worm screw and main nut are reground, thrust bearings reground, rocker shaft reground, new rocker shaft Phosphor Bronze bushes fitted and reamed to size, new ball bearings fitted, new lower oil seal and gaskets fitted, thrust bearing end float and rocker shaft mesh adjusted to original spec.

I also make and supply oil filter adaptor kits for the P4 Cyclops, and another adaptor kit for the P3 60 and 75 and Land Rover Series 1. These kits are made in solid aluminium and use a modern spin-on oil filter, replacing hard to get canister type oil filters.

Please contact me for further information, mobile 07966 136 890 or email dave.dt@frankdudley.com

Dave Dudley-Toole ***July 2012***

THE SPARE WHEEL – DON'T LET IT GET STUCK!

We're not sure if this tip's been around for as long as leaf springs, but it avoids being caught out by a common P4 pitfall.

Unless we're obsessed with originality, most of us fit radial tyres to our Rovers. And the chances are, we'll choose a slightly wider size than the cross-plies that the car wore when new; the most popular choice seems to be 185x15. Fine for roadholding and long life, but too wide to squeeze into the under-boot spare wheel holder. That means losing half the boot space to carry a spare wheel – not good anyway if you also want to carry anything clean next to it.

One solution is to carry a 165x15 – which should fit into the holder – as an emergency spare. Thanks to this being a popular choice for the 'real' VW Beetle and some older Saabs, the 165x15 can be easier to find as a car-spec tyre than the wider size, which many owners buy as 5-ply van spec instead. Of course it's illegal to run with one tyre narrower than the others unless it's a 'space saver' spare, which a 165x15 would be in the circumstances. That means marking the spare wheel as an emergency spare, with a 50mph/80km/h speed limit.

Even then, in most P4s it's still necessary to widen the spare wheel slot slightly. A small trolley jack and piece of timber is all it needs to raise the boot floor a smidgen.

Now comes the pitfall. Can you remove the spare? If it was a tight squeeze into the holder, you probably won't be able to. Perhaps not on the drive, and certainly not on a cold, damp winter evening when you can't afford to be late. And that's before a load in the boot presses the floor back to the old profile, before the trolley jack re-shaping.

The precaution to avoid this problem is shown below. Simply thread wire through the spare wheel's holes before pressing it into the holder, leaving enough spare wire to form a handle just inside the door.

We still recommend taking the spare out from time to time and putting it back into place. Firstly this enables the owner to check the pressure and condition. Secondly, it reduces the risk of the rubber 'sticking' to the metal and not wanting to slide out. If you've ever run a French car with the spare clamped in place under the boot floor, you'll understand the latter problem.

RELEASING THE BONNET CATCH WHEN THE CABLE BREAKS

Although I have owned two P4s (a 110 and an 80) for a few years, I don't claim to be an expert on them. However I was disturbed to discover from the manual that if the bonnet release cable was ever to snap, it would be impossible to open the bonnet!

I decided to see if there was another way to open it and there certainly is.

You need a slim shafted screwdriver, about eight inches long. Slide it at a steep upwards angle, behind the centre vertical radiator grille pillar, from the left side of the car towards the right, until it touches the bonnet release lever. A sharp tap will open it easily.

It would be better to try it before the cable actually breaks, with the bonnet open. Make sure the catch is kept well-greased.

I have heard of people destroying the radiator grille when trying to get to the lever, but this isn't necessary.

Ken Glass — ***January 2013***

(Editor's note: On earlier models there's a circular hole about 8mm diameter in the top of the front grille, to allow a screwdriver to be used for

this purpose. A number of members with later models claim that the hole isn't present on their cars. Opinions differ as to when the hole disappeared, although maybe a large proportion of P4s don't wear their original front grilles)

HOT STARTING PROBLEMS – MORE THEORIES

Those were two interesting letters in the last Overdrive. If it helps, I had a couple of pointers having also experienced poor starting when hot.

Like many others I exchanged leads, coils and rebuilt the distributor, including new cap and points. I even borrowed the carburettor from a fellow member's car, but all to no avail. I still felt it was an ignition problem as when it did eventually start it didn't splutter, but whatever the cause I still had to let it cool down. I decided to see if I could buy those connectors you attach between the plug lead and plug, to see if the spark was regular, particularly in size. Looking on the Internet I found an even better tool for connecting in series into the plug lead. This could be adjusted whilst running the engine, to see the strength of your spark.

This duly arrived and on start-up it showed a strong spark about 10mm in length, but as the engine increased in temperature I had to reduce size of the gap to keep the plugs sparking. I suppose it went down to about 2mm, the size of the plug gap.

Now when we tried to start the engine, there was no spark at all. However each time after the engine had cooled it would start but I found the reduction in spark strength was the same for all the plug leads, but interestingly no dropping occurred in spark size from the supply lead from the coil to the distributor.

It was looking very much like an HT problem to me now. Accepting that I had already changed the leads, coil and distributor cap, the only thing left was to try a new rotor arm which, in my ignorance, I had considered to be serviceable as it was an original.

I called the Distributor Doctor for a replacement rotor arm and helpful chat. It appears that in the 1960s the rotor arms were not of the earlier quality. The insulation properties were not as good and the brass pin which helps secure the rotating metal plate on the top was slightly longer and it was found that this was occasionally shorting out to the rotating spindle in the distributor.

I think the Distributor Doctor later wrote an article for Overdrive giving a useful insight into possible ignition problems.

A new rotor arm led to an immediate, positive response. I was on a high for a couple of years. It is always nice to find the problem rather than 'Let's try this.'

However, hot or cold, my engine was then again taking longer to start. The engine turned freely but was taking time to fire come to life, again without any spluttering. I confess that at this point a bit of luck went a long way. I decided that rather than keep getting in and out of the car, I would switch on the ignition and short out the main terminals on the starter coil. Suddenly the engine started immediately. Again I cross checked with the ignition switch, a slow response but eventually it started. Back to shorting out the starter coil and an immediate start. Just imagine without luck I could have been chasing fuel or ignition problems. The sun shines on the righteous!

I think a fault had occurred in the coil, although turning the engine so much left little battery power to feed the ignition circuit to start the engine. With a new coil I am back on a high and hanging in there.

Check eBay for adjustable spark gap testers, but beware of high postage costs, particularly from the USA.

Norman Hills

I searched through the letters and comments in Overdrive 210 about poor hot starting and nowhere did I find any reference to Rover's own cure for the problem, that became a standard for its carburettor cars from P6 onwards.

When my Rover 90 was being used frequently (This is a whole other story that I might have news about later this year) I often had problems of poor hot starting. I would nip into a newsagents or refuel or something, and when I returned the car would take ages to settle down to even running, with revs being up and down, making the car undrivable for several minutes. It also often took ages to actually get it started.

I have forgotten now how I came to try this fix, but it was an instant and

100% successful cure. I put an insulator block from a P6 between the carburettor and the engine head. This pushed the carburettor out towards the bonnet on my 90, but it just fitted. This block had the effect of insulating the carburettor from heat soak from the engine, with its unwanted side-effect of vapourising fuel. It was an instant fix, I never again had any problems with this alteration except in the extreme cold weather of 1996/7 when the insulation worked too well, keeping the carburettor too cold, and reducing its ability to vapourise fuel, thus reducing top speed to about 30mph – it was a harsh winter, and frankly I couldn't safely drive at much faster, but that isn't the point. (Classic Cars, March 1997, p157)

So: fit an insulator block from a P6. I remember I had to carve some of this black plastic away around the opening to fit over a flange, but otherwise everything worked perfectly, even the bolts were just long enough.

Matt White ***Overdrive Editor 1980-2005***

With regard to the above problems, although my P4 does not suffer from this problem (touch wood), I have found the following procedure usually works. Over the years and with several different cars it has been successful. Slowly press the accelerator right down to the floor, engage the starter. Slowly to avoid the accelerator pump flooding the carburettor. This presumably weakens the mixture. I cannot see that an electrical fault can be at the route of the problem.

Jim Russell

I read with interest the piece in previous issues about poor hot starting, and would like to know what is the correct running temperature for the engine of my 1964 P4 110.

I do not have problems with hot starting, and taking a temperature of engine with an infrared thermometer pointing at the block just behind and to the rear of the dynamo it registered 77°C in normal running temperature.

My temperature gauge always shows just at the top of the N, never getting any hotter even in traffic queues, so is my gauge reading wrong or are there some items I should check?

As regards fuel vaporisation I had a problem last year, going to the Goodwood Revival in my 1964 Humber Super Snipe (sorry – the Rover went the previous year!) We were in a massive hold-up on the A27 for about one and a half hours, just crawling along and finally, less than half a mile from the turn off her engine petered out.

After my wife and three grandchildren (they were all 18 to 20 year olds except for the wife) pushed us to the side of the road, where we sat there with the bonnet up. After ten minutes she started first time with no further problems. The temperature gauge was reading slightly above normal but nothing to worry about, so I put this down to fuel vaporisation.

I have been told that the unleaded fuel we use nowadays does tend to run hotter – is this correct?

D D Sutherland

CLOTH HEADLINING RESTORATION ON A 1955 P4 90

I waved goodbye to my trusty 1969 Triumph, having decided to look for something bigger, older and with four doors.

When browsing through various websites I saw the Rover P4 and thought "I quite fancy one of those if I can find one in good condition at the right price."

On browsing the classic car adverts (a daily exercise) I saw a new entry from Northern Ireland; it was a 1955 90 registration number SZ 1523. I e-mailed the seller, who informed me that he was selling the car for a friend who was poorly. The car had just passed its MoT in Ireland, which is far stricter than the one in England. From the photographs, the body looked in good condition (they can give a false impression) but the carpets and headlining were poor.

A friend who lives in Northern Ireland kindly went to have a look at the car for us and confirmed that the details were true and also said that the engine ran sweetly. Decision: should I go to NI or trust my friend?

I decided to trust my friend, go ahead and purchase the car unseen by me. Unfortunately the owner of the car then passed away, so arrangements for the purchase of the car had to be postponed for a little while until the family was available to make money exchanges. Sending a cheque to someone you have never met is all down to pure trust. All went well, documentation was all returned to me and arrangements were made to have the car shipped over to Stone in Staffordshire. "Suzy" the Rover then came to England on a big transporter which was to big to come to the house, so she had to be delivered to our friendly local garage.

On meeting Suzy for the first time, my wife Norma and I were delighted and we have remained so. Suzy's outside was indeed very good, but on inspecting the inside we found that moths had had a whale of a time with the carpets and headlining.

Enter Norma who does not drive, but loves older cars and who is also a dab hand at sewing. "I could make one of those", she said after glancing at the tatty headlining. "You can save me the old one for a pattern, I have two heavy-duty curtain liners which I am sure will do the job". I set about removing the side pieces and front screen headlinings, then realised that I would not be able to take the frame and head lining out without removing the front bench seat After this was removed it was a doddle. The frame and lining came out through the front door. On inspection, the headlining was clipped into the frame and held through seams onto the frame's steel bars.

Together, Norma and I removed the old lining from the frame. Norma then set about using the remains as a pattern, and using the curtain linings double, leaving extra material around the edges to make runners for the frame and enough on the sides to be clipped back onto the frame.

For the front and sides Norma was able to use the boards as patterns. The longest part of the job was stitching the seams to the frame. This needed to be done by hand, after the frame itself had been cleaned and painted.

The interior of the roof panel and the headlining frame were both covered in rust, which I sanded down and painted first with rust converter and then with two coats of paint.

The main headlining panel then clipped onto the frame and was now ready to be re-installed into the car, with much pushing and shoving by the two of us. We managed to screw the frame back into the car using only eight small screws but it took four long hours, working in a cramped position. We then replaced the front, side and back trims, fixed the lights, switches and straps back in, and sat back to admire our handiwork.

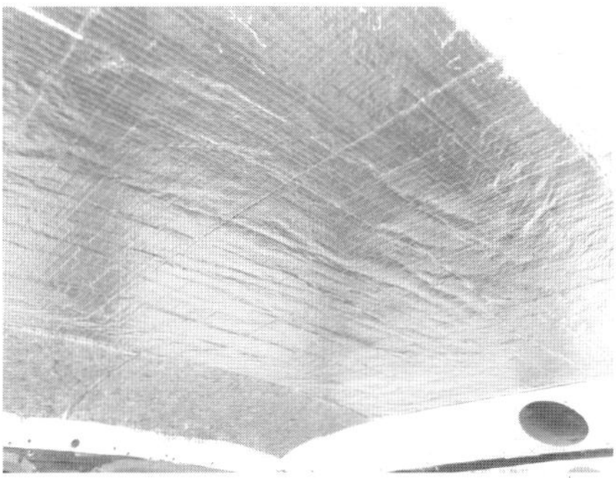

I estimate the time taken to be about ten hours in total. You don't have to spend a fortune to improve your car.

Tony Lockett ***September 2013***

P3 AND P4 STEERING BOXES

The steering box on your P3 or P4 is probably one of the most neglected, but vital parts on your car. The reason it's so neglected? Simple, its position. On the P4 it's buried beneath the carburettor and brake servo, making it very difficult to get at. P3s are much easier to get at and remove, but they are still a very neglected item. Your steering box deals with tremendous stresses and strains every time you drive your treasured possession, but have you any idea of its internal condition?

P3 and P4 steering boxes are very similar in construction, both being Burman re-circulating ball type, but the P3 box is quite a bit bigger and stronger than the P4's. However they both wear out in very similar ways.

Most if not all boxes leak like sieves. We've all been there – you put oil in the top, and within minutes, there's a puddle on the floor!! P3s have a 'proper' lip type oil seal in the bottom of the drop tube, but yours is probably over 60 years old and will almost certainly have gone very brittle and will leak. In the P4 the 'seal' for the forked drop arm is just an 'O' ring, made originally from pure rubber that over the years usually becomes very hard and brittle. With time, the 'O' ring changes to a 'D' shape, i.e. a flat develops on the inside which creates a gap for the oil to leak through. It looks pretty easy to replace the 'O'` ring in situ by removing the drop arm, forked drop arm and top cover. It is certainly possible to do all this and to lever out the old 'O' ring, but what owners don't realise is that on top of the old one there is a thin, flat washer. When the end cap is pressed on to the drop tube, this washer compresses the 'O' ring, keeping it firmly in position. It also compresses the 'O' ring to grip the forked drop arm shaft more tightly, preventing oil leaks. So, after many hours of head scratching and toil, you've finally got the new `O` ring in position, and stopped the leak, no more oily puddles on the garage floor. Brilliant!! But wait. When you took the forked drop arm out, did you notice how badly worn it was at the top and the bottom? Your forked drop arm has now become oval instead of round, and so have the upper and lower, thin-walled phosphor bronze bushes in the drop tube. You now face the age old problem of 'What do I do about that???' The usual answer is 'I know, I'll screw the top adjuster down to take up the slack!' There's no doubt that this action will take up the slack, but you've just brought yourself more problems by putting even more pressure on other badly worn parts, and you've also made your steering even heavier!!!

On P3 models, the steering box casing and drop tube are a one-piece aluminium casting (P4s have a tubular steel drop tube), and the P3 drop tube is machined to accept the forked drop arm. When the drop tube and forked drop arms wear out, the only way to rectify the wear is to grind the forked drop arm and bore the casing oversize at both ends, then fit thin-walled phosphor bronze bushes and line ream to size.

On the P3 and P4 the worm screw, main nut, thrust bearings and forked drop arm are the most crucial parts of your steering box. They are also the parts that suffer most wear and damage. Most owners think that there is only one means of adjustment on the box, at the adjusting screw on the top cover. There are in fact two adjustments. The other means of adjustment is by shimming the end float in the worm screw thrust bearings, but there's no point trying to adjust the end float if the thrust bearings and ball bearings are worn or damaged, which THEY ARE!! Thrust bearings always get damaged or wear out from the centre in a downward direction, this is due to excessive adjustment of the top adjusting screw. The top adjuster should only be used to reduce lift in the forked drop arm, thereby increasing the mesh with the main nut. But this is purely academic if your worm screw, main nut, thrust bearings and the ball bearings are worn. All you will do is make your box stiffer, wear out even quicker and cause MORE DAMAGE!!

I've been fully reconditioning P3 and P4 steering boxes for over five years, but I'm still totally shocked at the internal condition of many, many boxes. It is truly amazing how the cars ever steer at all. The scrap parts rate is frightening: worm screws 40%, main nuts 30%, thrust bearings 50%, casings 15%, forked drop arms 20% and 'O' rings 100%. Casings and forked drop arms can crack due to impact damage, but some forked drop arms crack because of excessive over tightening of the top adjusting screw, DON'T DO IT!!

The parts in the attached photos were all taken from cars that were MoTd and on the road. Very, very frightening!!

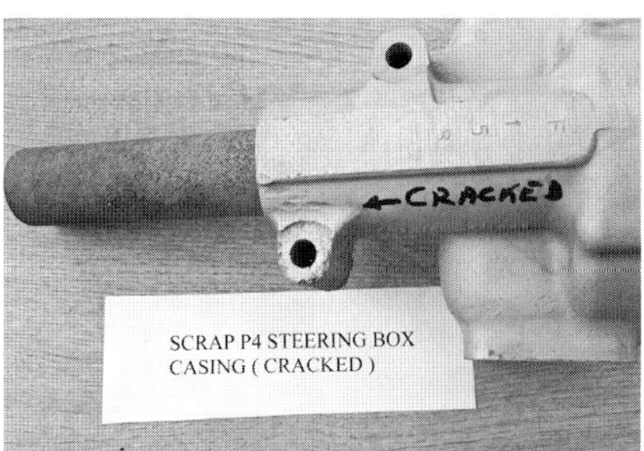

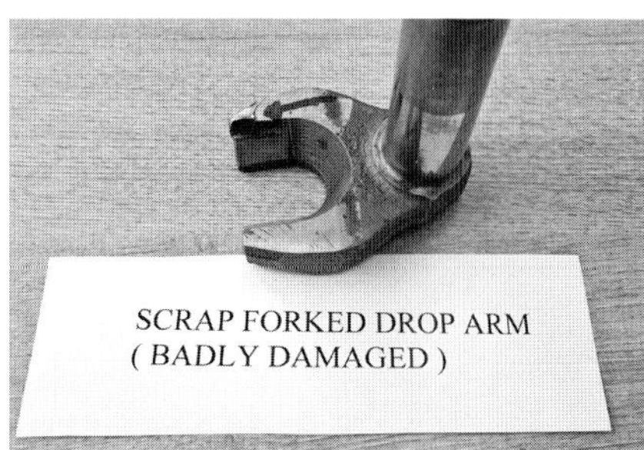

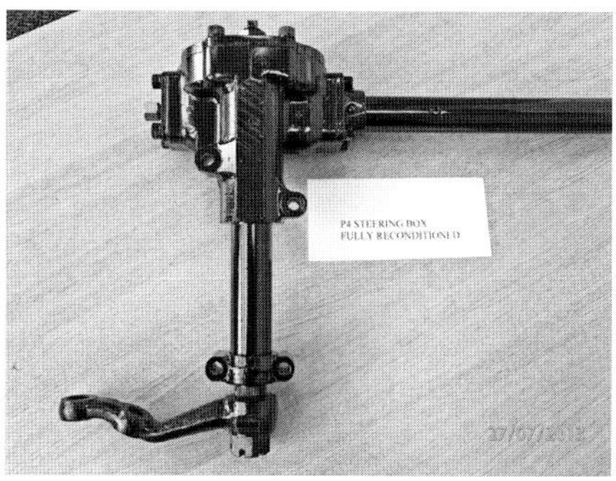

I can fully recondition all P3 and P4 steering boxes on an exchange basis, usually from stock. All of my steering boxes are fully stripped, cleaned and crack tested. The worm screw, main nut, forked drop arm and thrust bearings are re-ground. New phosphor bronze bushes are fitted to the drop tube and reamed to size. All new ball bearings are fitted to the thrust bearings and main nut, the thrust bearings are shimmed for correct end float and the main nut is adjusted for correct lift. All new gaskets and a new 'O' ring fitted (P4) (lip seal on P3), and finally the rebuilt box is painted in high gloss black.

I can also recondition left hand drive boxes, but because of their rarity in the UK I would need the old one first to recondition.

I recently managed to get hold of a P2 steering box and hope to get it apart quite soon with a view to seeing if I can also recondition these. Hopefully I will have news on this shortly.

I can also recondition steering idlers on an exchange basis and supply a new oil filter adaptor kits for early P4s (cylinder head type), utilising a modern screw-on filter, and a canister type oil filter adaptor kit for P3 60/75 and Series 1 Land-Rovers, again utilising a modern spin on oil filter.

Please feel free to contact me at any time for help or advice regarding your steering, or oil filter, my email address is dave.dt@frankdudley.com, or call me on 07966 136 890.

Dave Dudley-Toole

INVESTIGATING P4 FUEL GAUGE SYSTEMS

Before proceeding, the first point to make is one of safety. There is an inherent risk in making measurements on in situ fuel sender units. These incorporate a float and variable resistance, the whole assembly being screwed into the off-side of the petrol tank. All is sealed in place by a gasket but, over the years, seepage could be taking place. Unwanted sparks and resultant petrol vapour constitute a dangerous combination! The risk of sparks can be eliminated by ensuring the ignition switch is off but it is even better to disconnect the battery itself.

The next point to note is that the fuel gauge system is different for each range of cars and the individual items are not interchangeable. For the 95 and 110 models, the system also includes a voltage regulator behind the instrument panel (not to be confused with the dynamo regulator on the engine bay bulkhead). The appropriate part numbers are as follows:

Sender unit for early cars – 235818
Gauge for early cars – 236652

Sender unit for later cars – 536036
Gauge for later cars – 536034
10 volt regulator – 531651

If there is any doubt as to which system is fitted to your car, identification is easy: with the early gauges, when the car is in motion, the pointer will wag like the tail of a dog that has been offered a titbit – very irritating to some, but part of the car's charm to others!

Although the gauge systems are not the same, in one respect they are similar: each relies on a float which rides on the fuel in the tank and varies a resistance according to level. The resistance controls the current through the fuel gauge.

If you wish to change the polarity of the car, for either system, there is no need to alter any of the connections. However, if you have a 110 or 95 which has been retrofitted with a modern fuel gauge voltage regulator, it will be necessary to change this item if you decide to opt for negative earth.

The fuel gauge sender unit is accessed by removing the off-side rear wheel.

The manual suggests fault-finding by substitution or by continuity testing. These days, cheap, accurate multimeters, available from stores such as Maplins, are a useful addition to the fault-finding armoury. A suitable meter from this company has a part number of N20AX and is priced at about £10. Very useful too would be a pair of clip-on probes; these cost around £5 and bear a part number of N21CB.

FUEL GAUGE SYSTEM FOR THE EARLY CARS

As stated, the underlying principle of operation for the early cars is the same as for the later ones. A sender unit in the petrol tank incorporates a variable resistance attached to a float. The float moves according to the fuel level in the tank, causing the resistance to change. The resistance allows a current to flow to the gauge which is related to the fuel level. When the tank is empty, the current through the gauge is at a maximum and, when measuring full, at a minimum. (This is the opposite situation for gauges on the later cars).

The gauge exploits the electromagnetic motor principle, namely that if you pass an electric current through a coil, a magnetic field results that can be used to produce a turning effect. The instrument can be described as a moving iron meter since the pointer is attached to a piece of soft iron plate: this is pivoted between two electromagnets which comprise of bobbins. The bobbins contain several coils of fine wire. An energising current flows through the coils and the magnetic field produced causes the plate to move and thus the attached pointer to deflect. The soft iron plate is asymmetrical in shape and acts like a flywheel so that, with the ignition switch off and no current flowing, gravity forces the heavier section downwards and the pointer registers empty.

Unlike the gauges in the later cars, a voltage regulator is not required. The design cleverly incorporates the two separate coil windings and each one is affected similarly by any change in battery voltage, so self-regulation occurs. The situation is analogous to a tug-of-war where, before the contest starts, the initial balance is still maintained if a person is added to each team (voltage increased) or removed from each team (voltage decreased).

Faults with the system are now discussed, but before going into detail, it is worthwhile carrying out a simple test. If you suspect a problem with petrol measurement then, using the changeover switch on the instrument panel, switch the gauge to register engine oil level. A reasonable reading, which can be confirmed by the dipstick, indicates a good fuel gauge. (If the reading is "strange" this doesn't necessarily mean a defective gauge: it could be due to a faulty changeover switch, a wiring problem or the sender in the sump misbehaving. More investigation is needed).

Prior to considering faults and their symptoms, it is relevant to ask the question: after all these years, do we still have the right sender in place and how would we know? With a full tank and an incorrect sender unit, the gauge will measure around ¼ and, as the tank empties, the pointer will nudge disconcertingly towards 'maximum'!

SYMPTONS REMEDIES AND MEASUREMENTS

Gauge reads zero regardless of tank contents

Is a voltage arriving at the 'B' terminal (B for battery) of the fuel gauge?

Looking at the front of the gauge, the 'B' terminal is the one on the left-hand side. With the ignition switch on, apply one probe of your multimeter to the terminal and the other to the body of the gauge or some other convenient earth point. You should read around 12.6 volts (a typical value for a fully charged battery). If you see 0 volts then check the fuse between A3 and A4. If the fuse is OK, trace the wiring from A4 to the gauge. It is a green wire that goes to a bullet splitter connector behind the instrument panel and then to the petrol reserve switch, before arriving at the gauge. (Please refer to wiring diagram).

Is the earth lost on the fuel gauge? It is not obvious how the gauge is earthed! On the back of the instrument, there are two terminals, one labelled 'B' for battery (already mentioned) and the other 'T' for tank. There is a further, vital, connection – the body of the gauge. It must be earthed. Earthing is via the gauge locating studs and continued by an earth strap to the back of the binnacles (normally a black wire). Check that the gauge is bolted firmly to the mounting studs and that the strap wire is making a good connection.

Are the 'B' and 'T' connections interchanged? If this is the case, the gauge windings will probably have burnt out rendering it inoperative: it will show 'empty' only. A new or second hand gauge will be required. (To check the gauge, please refer to 'Gauge Measurements' under Appendix A).

Is the 'hot' connection on the sender or that to the 'T' terminal on the gauge lost? On the sender unit, you will find two terminals, one is connected to an earth point on the chassis and the other, a green/black wire, finds its way to the 'T' terminal via the petrol/oil changeover switch. Check that the connectors are tightly attached to their terminal posts and that nothing is loose on the back of the switch.

The gauge could be reading 'empty' because of mechanical failure? To check this possibility, it must be removed. Undo the two 4BA securing nuts and lift the gauge from its studs. Extreme care is required to avoid bending the pointer as you lift the instrument from the back of the binnacle. If all is well mechanically, the pointer will move to the 'full' position as you gently rotate the gauge through 90 degrees, such that the 'full' gradation ends up downwards. If you have removed the gauge, this is a good moment to check it electrically – please see Appendix A, 'Gauge Measurements'.

The sender unit float could be stuck in its lowest position? If the tank is more than a quarter full, this situation can be verified by measuring the resistance between the sender terminals. Before measurement, turn the ignition switch off or isolate the battery. Now remove the connecting wires from the terminal posts and, with the multimeter set to 200 Ω (ohms), measure the resistance between the posts. If the float is jammed on 'empty', you will see a value of around 5Ω or less.

To remedy matters or further investigate, the sender must be unscrewed and removed from the petrol tank. At the same time, measurements can be made. Please see Appendix A, 'Removing the Sender, Remedying and Measurements'.

Gauge reads full, regardless of tank contents

Has the sender lost its earth? The electrical circuit between the sender and the fuel gauge is completed by an earth connection at the sender end. A black earth wire leaves one of the terminal posts on the sender and is bonded to the chassis forward of the offside rear wheel. Check the terminal post connection and the chassis bonding point. The area where the wire is attached to the chassis is often compromised electrically by mud and debris thrown up from the rear wheel. Here, surface rusting is likely too. If you suspect trouble, remove the self tapping screw; scrape the area where contact is made so it is really clean and rust free; remake the connection and apply waxoyl liberally. (It may be easier to cut the earth wire loose; crimp on another eyelet; drill a new hole close by and remake the connection. Preferably use a stainless steel self-tapper and shakeproof washer).

Is the gauge mechanically damaged such that pointer deflection is permanently at 'full'? This situation can be verified by removal of the gauge, as described previously. If all is well, when the gauge is gently rocked from side to side, the pointer will deflect accordingly. Stickiness means replacement! (Assuming the pointer moves smoothly, if not already done, this is a convenient time to carry out an electrical check – please see under 'Gauge Measurements', Appendix A).

Is the variable resistance on the sender faulty? The resistance can be checked by measurement. Before proceeding, ensure that the ignition switch is off or the battery is isolated. The resistance range is between about 5Ω (empty) and about 90Ω (full). To make a reading, switch the multimeter to its 200Ω range which is the most convenient one. Remove the wires from the sender terminal posts and place the probes across the terminals. If the multimeter measures much more than 90Ω then here lies the problem. Please refer to Appendix A, 'Removing the Sender, Remedying and Measurements'

Gauge readings are patently inaccurate

An earthing problem? Poor earthing may be the reason. In particular, check the sender earth connection on the chassis; as mentioned previously, its location is very vulnerable. The resistance across this connection should be 0Ω but it may be several ohms higher due to the presence of corrosion and dirt.

Is it the sender unit? Another source of inaccuracy is the variable resistance part of the sender unit which can malfunction. Please see 'Removing the Sender, Remedying and Measurements' to analyse further.

FUEL GAUGE SYSTEMS FOR THE LATER CARS

A description of how these gauges work is given in the Workshop Manual. For convenience, my own brief description is given here.

Basically, the gauge receives a DC current which is related to the position of the float belonging to the petrol or oil sender unit: which one depends on selection by the instrument panel changeover switch. The higher the float, the larger the current supplied to the gauge and the greater its pointer deflects. Thus, 'full' relates to maximum current, and 'empty' to minimum current.

The gauge itself is known as a hot-wire ammeter, that is to say the current passing through it causes a heating effect and the heat generated makes the fuel gauge pointer move.

In the interests of accuracy, the fuel gauge is fed via a 10 volt regulator. (This regulator is not to be confused with the one controlling the dynamo. The latter is found on the engine bay bulkhead).

The 10 volt regulator is located at the back of the instrument panel and also works on a heating principle: fed with the vehicle's 12 volt supply, it switches on and off rapidly so that the average voltage applied to the gauge is 10 volts. (An analogous situation would be to switch a lamp in your house on and off very rapidly: the lamp would still illuminate, but dimly, because the average voltage applied would now be less than the nominal 240 volts). If the battery voltage rises, the switching speed of the regulator adjusts so the regulator maintains the required 10 volt output.

Since both the regulator and fuel gauge work on a heating principle, they are not polarity conscious: if the vehicle undergoes a polarity change, then connections to these items remain the same. (But, if a modern solid state regulator has been fitted, this will need to be substituted).

SYMPTOMS, REMEDIES AND MEASUREMENTS

Gauge reads zero regardless of tank contents

Is the temperature gauge working? Warm the engine up and look for a reading on the temperature gauge. If the temperature gauge is operating, then this means that the voltage regulator is functioning because it feeds both gauges. (If the temperature gauge fails to read, it could also be due to a faulty gauge or temperature sender unit – Sod's Law)!

Is the voltage regulator receiving 12 volts? If the temperature gauge is not responding, then our investigation must centre on the voltage regulator. It receives a 12 volt supply from the fuse between A3 and A4. Check this fuse.

If the fuse is sound, with the multimeter set to DC voltage and an appropriate range selected (say 20 Volts), look for a reading of about 12.6V (charged battery) on the 'B' terminal of the voltage regulator. ('B' stands for battery). When making this measurement, the positive probe of the meter should touch the regulator body (which is earthed) and the negative probe should contact the 'B' terminal. Reverse these connections if the car has been changed to negative earth.

Is there a wiring problem? If the A4 fuse is OK, but there is no voltage at the 'B' terminal, this suggests a wiring fault or that the earthing to the voltage regulator has been lost. Trace the 'hot' wire from A4: it is green and feeds a multiway bullet connector before arriving at the 'B' terminal. The regulator is earthed by its mounting studs and through a black earth wire at the rear of the binnacles. Check that the stud nuts are tight and that the earth wire has not come loose.

Has the voltage regulator failed? If a reading of around 12.6V on the 'B' terminal is obtained, this suggests that the regulator has died. Death can be confirmed by moving the multimeter probe from the 'B' to the 'I' terminal – the meter will show zero volts. ('I' stands for instrument). Please see Appendix B, 'Testing an original regulator'.

Is the regulator connected correctly? At this point it is worth noting that the connections on the regulator cannot be interchanged: 'B' must go to the battery via the A4 fuse and 'I' must go the instrument gauge.

(Replacement electronic regulators are now available and are far superior to the original bi-metal switching type. However, the replacements are polarity conscious and if connected wrongly will be destroyed immediately! If you are contemplating buying an electronic regulator, state the polarity of the vehicle to the supplier and, when fitting, make sure you connect 'B' and 'I' appropriately).

Has the fuel gauge failed? If the regulator is not defective, attention must be turned to the gauge or the sender unit. A quick check on the gauge can be made by selecting 'oil' level. If a reading is not obtained, then the gauge is suspect, (although the reason could be a faulty selector switch, a wiring fault or the oil sender unit itself – Sod's Law again)! Electrically, the gauge can be checked by measuring its resistance. To do this, disconnect the wires from the back and, with the meter set to a range of about 200Ω (ohms), apply the probes to the terminals. A reading in the region of 60Ω ohms will be obtained for a good gauge. It is possible that the gauge movement has failed mechanically and this would be indicated by the needle sitting lopsided across the dial face.

Is the tank sender unit faulty? If the gauge is OK, the fuel sender unit must now be examined. The multimeter can be used for checking this item. In the interests of safety, first ensure that the ignition switch is off or the battery is disconnected. Remove the earth and hot wires from the unit. Set the multimeter to a range of about 2000 Ω (ohms), sometimes designated as 2KΩ and apply a probe to each connector lug. A reading in the range of between 250Ω and 20Ω should be obtained. (These figures represent tank 'empty' or tank 'full' respectively. 10% either side of the figures is quite acceptable). It follows that if the tank contains some fuel, then an intermediate value, within this range, will be indicated.

If no reading results, then the sender is faulty and needs to be removed from the tank for further investigation. Please see Appendix B for procedures and details.

If you are considering buying a second hand sender, do make sure it is for a later car (95 or 110): as stated previously, the early cars had a different sender unit which is not compatible. The one required has part number 536036.

For a full tank, the gauge reads less than it should

Is the regulator faulty? The regulator is designed to stabilise at 10 volts. If it settles at a lower value then the whole range of current through the fuel gauge will be reduced, giving lower readings throughout: this will be most obvious when the tank is full. Please refer to 'Testing an original regulator', Appendix B.

Is there a wiring problem? A poor earth or bad connections will also reduce the range of current to the gauge with similar effects to a regulator fault. In particular, it is well worth checking the sender connections as these are exposed to the elements.

Is the gauge faulty? With time, the gauge characteristics may have changed. I measured two and found significant differences and would suggest changing "if all else fails".

For a full tank, the gauge 'pins' beyond the full mark

Is this another regulator fault? With the original regulator, when the ignition switch is turned on, the fuel gauge should move initially beyond the full mark and then fall back. This behaviour is normal and caused by the regulator needing time to adjust to 10 volts. Before beginning to operate, it will apply 12.6 volts (normal value for a fully charged battery) to the gauge. The raised voltage produces more current that usual and, momentarily, causes a greater needle deflection. If the gauge continues to over-read, it is possible that the regulator contacts have shorted out or the switching action has ceased: in either case, the battery voltage is fed uninterrupted to the gauge. (But, a modern sender will cause an over-read too! – see Appendix B, 'Measurements').

Is the gauge suspect? As mentioned above, the gauge parameters may have changed over the years so that the instrument reads high.

Are the readings on the gauge patently inaccurate?

Is the sender misbehaving? This is the most likely cause of the problem.

The float may be sticking so the change in resistance is not smooth. Equally, the wiper arm could be making intermittent contact with the resistance. The best way of confirming culpability is to remove the sender and test it on the bench, as detailed under Appendix B.

Has the wrong sender been fitted? It will not necessarily be obvious if an early sender has been fitted: if there is some fuel in the tank, the gauge will stabilise somewhere in the middle of its range. But, as you drive the car, the pointer will deviate towards 'full' as the tank empties!

Peter Nairne

APPENDIX A – EARLY CARS

Removing the Sender, Remedying and Measurements

Before removal, ensure the tank is less than a quarter full (there is a drain plug on the underside). When you withdraw the sender unit, be careful not to bend the float or its support rod. Once out, check to see if the float is perforated and contains fuel: if so, this will cause the gauge to register 'empty' continuously.

Assuming the float is sound, it is possible that the float pivot bearings are sticky. To find out, support the sender unit on its side in a soft-jawed vice and try moving the float between its extremes. It should be free enough such that blowing on the float causes movement. If stiffness is encountered, spray the pivot bearings with WD40 then move the float several times between the limit stops until it frees up.

With the cover removed, examination will reveal a wiper arm operated by the float. The wiper slides along a tightly wound coil of resistance wire and picks off a resistance that corresponds to the level in the tank. Sometimes the wiper makes poor contact with the resistance wire, creating an accuracy problem. The contact surfaces can be cleaned by inserting a piece of card between the contact point on the arm and the resistance wire, then rubbing the card to and fro. Additionally, whilst holding the inserted card, move the wiper arm and card several times along the resistance. This action will clean the contact points on the variable resistance.

It may be beneficial to re-tension the wiper arm slightly by carefully bending it inwards to increase the contact pressure.

Since the sender is on the bench, it is a good idea to do some electrical tests – please see typical results below.

Sender Measurements

These measurements are made with the sender unit held in a soft-jawed vice so that the float arm can be tracked gently from one end to the other. As the arm is slowly moved, the resistance should change smoothly: severe jerkiness implies intermittent loss of contact with the wiper arm or wear on the resistance wire.

For convenience, the multimeter is set to 200Ω. Apply the probes to the terminal posts, preferably with croc clips, and then move the float across its range.

My results are shown below with the corresponding currents that would be fed to the fuel gauge. Please note that your measurements may vary by as much as 20% due to manufacturing tolerances.

Fuel in tank	Resistance	Current through gauge
Full	92Ω	110 mA*
¾ Full	57Ω	120 mA
½ Full	40Ω	130 mA
¼ Full	19Ω	150 mA
Empty	0Ω	190 mA

* 1mA = 1/1000 Amp

Gauge Measurements

Terminal B (battery) to Terminal T (tank)	60.6Ω
Terminal B to Body of Gauge	157Ω
Terminal T to Body of Gauge	98Ω

APPENDIX B – LATE CARS

Testing an original regulator

If measurement is made on an original regulator which is functioning, i.e. with the probes on the "I" terminal and the regulator case respectively, then the multimeter reading will fluctuate rapidly from about 12.6V (battery voltage) to a lower value. Theoretically, the meter should measure 12.6V and 0V in quick succession as the regulator switches on and off. But, being digital, the meter is unable to react quickly enough to register 0V and thus shows a voltage variation between the nominal battery voltage of 12.6 volts and some lower value (around 7 volts). Note: these regulators do not start instantaneously when voltage is applied to the 'B' terminal – several seconds are required before the heating effect takes place and the switching action begins: thus, initially, the full battery voltage will appear on the "I" terminal and the fuel gauge will read high.

Testing Sender Units – original and replacement

Before extracting the sender, recheck that the ignition switch is turned off or the battery is isolated. It is also advisable to drain the tank down so that it is about a quarter full (a drain plug is located on the bottom of the tank). When you withdraw the sender unit, complete with float, take care not to bend the float support rod or damage the float on the end of it.

Once extracted, the sender unit can be held sideways in a soft-jawed vice for testing but before doing so, ensure that the float is not perforated and fuel has not entered it. Held sideways, the float can be moved between its limits. Does it move smoothly from one end to the other? (If you blow on the float, your breath should cause some movement). If the float is sticking, apply WD40 to the pivots and rock it gently to and fro between the limit stops.

If the float moves freely, now is the moment to unscrew the self-tapping cover screws and take off the cover plate from the resistor box. You will see a resistance comprising of resistance wire wound tightly on a tapered former. Bearing on this wire, is a wiper arm which slides along the former length: its position, at any given time, is determined by the height of the float. Thus, the wiper 'picks' off a resistance that corresponds to the float height; this, in turn, relates to the amount of fuel in the tank.

With the multimeter set to 2KΩ (2000 ohms) apply or preferably clip the probes across the sender terminals and very slowly move the float from one limit to the other. As previously mentioned, on a working sender, the multimeter will measure about 20Ω at one limit (tank full) and about 250Ω at the other limit (tank empty). As you move the float between the two limits, the reading should steadily change in value. A sudden, large change indicates that the wiper has lost electrical contact with the resistance.

If the unit does not check out satisfactorily, as above, there are three common causes:

1. The connector lugs are riveted to the unit and, with age, they come loose; this creates a poor or intermittent connection. Wobble the lugs to check for tightness. A loose connection can be made good by applying

solder around the rivet, after it has been thoroughly cleaned first. When soldering, ensure that the solder sets as a bright blob around the rivet and that the offending lug is now tight.

2. The contact track for the wiper, along the resistance wire, is dirty. To clean, take a thin piece of card and rub it along the track. Also, rub between the wiper and the track.

3 The slider has lost its tension and the pressure is insufficient to produce good contact with the resistance wire along all of its travel. Sometimes, it is possible to restore tension by carefully bending the end of the slider slightly inwards.

The following are my test results for a good sender unit:

Fuel in tank	**Resistance**	**Current through gauge**
Full	22.5Ω	120 mA*
¾ full	44.7Ω	88 mA
½ full	75Ω	70 mA
¼ full	121Ω	51 mA
Empty	248Ω	33 mA

* 1mA = 1/1000 Amp

Modern compatible sender units are available but connections to them are confusing because there are three terminal lugs, not two! The reason for this is that later versions have been designed to illuminate a ‘tank empty’ light as well as feed a fuel gauge.

Which is which? The earth lug is obvious because it is bonded to the external surface of the unit. The fuel gauge lug, the one we want, can be found by measurement. With the unit held sideways in a soft-jawed vice, move the float so that it is in the empty position. Set the multimeter to 200Ω (ohms), then with one probe on the earth lug and the other on one of the two remaining lugs, you will measure around 2Ω to 3Ω (ohms) for the warning light terminal: it’s the other one you need.

Here are my readings for a modern sender unit:

Fuel in tank	**Resistance**	**Current through gauge**
Full	9.5Ω	140mA
¾ full	49Ω	88mA
½ full	77Ω	70mA
¼ full	164Ω	44mA
Empty	241Ω	33mA

* 1mA = 1/1000 Amp

These readings confirm that a modern sender is reasonably compatible with an original, in particular, that the important 'tank empty' resistance is virtually the same between ancient and modern. However, at the other end of the scale, the modern unit has less than half the resistance of the original and that means that the gauge will register beyond 'full' with a brimmed tank - not too serious!

It is worth noting that neither sender will produce a true 'empty' indication because, at tank empty, there is still a tiny current flowing through the gauge (33mA): this causes a small deflection of the needle above zero! Thinks: "Is that why Rover included a reserve switch?"

Peter Nairne ***2012 Yearbook***